Springer Earth System Sciences

More information about this series at http://www.springer.com/series/10178

Jean-Pierre Peulvast · François Bétard

Landforms and Landscape Evolution of the Equatorial Margin of Northeast Brazil

An Overview

Springer

Jean-Pierre Peulvast
Université Paris-Sorbonne
Paris
France

François Bétard
Université Paris-Diderot
Paris
France

ISSN 2197-9596 ISSN 2197-960X (electronic)
Springer Earth System Sciences
ISBN 978-3-319-18202-5 ISBN 978-3-319-18203-2 (eBook)
DOI 10.1007/978-3-319-18203-2

Library of Congress Control Number: 2015938746

Springer Cham Heidelberg New York Dordrecht London

Printed on acid-free paper

Springer International Publishing AG Switzerland is part of Springer Science+Business Media
(www.springer.com)

Preface

This book is the result of more than 15 years of fieldwork in the Ceará and adjoining states (Rio Grande do Norte, Paraíba, Pernambuco, Piauí). Its preparation was undertaken in a period characterized by important progress in the knowledge of the geology and geophysics of Northeast Brazil and, more generally, of passive margins, basement areas, and sedimentary basins. New methods such as thermochronology also contributed to give new insights into topics such as denudation rates and long-term landform evolution models, raising up renewed discussions when confronted to more classical (morphostratigraphic) approaches. Having gathered a wide and original set of geomorphic, morphostructural, pedological, and other data from fieldwork and processing of various types of digital data (digital elevation models, satellite images…), the authors had published their results in many separate articles, but they finally felt the necessity of publishing an integrated work, in the form of what appears as the first and unique synthesis on the landforms and landscape evolution of Northeast Brazil.

More than a simple monograph, the authors present a comprehensive geomorphic overview of a large tropical region where they show how deciphering the long-term landform evolution helps in understanding the present set of landscapes and morphodynamic environments. Illustrated by original maps and more than 80 color photos and 3D images, it is centered on an original analysis of landforms and soils in a key area allowing the understanding of long-term to short term geomorphic evolution on the equatorial passive margin of Brazil. This part of the Brazilian "Nordeste" displays stratigraphic landmarks whose interpretation reveals the age and nature of landforms, leading to a reconstruction of the geomorphic history by the means of combined morphostratigraphic and morphopedological approaches. Beyond the role of differential erosion related to moderate post-oceanic opening uplift, the plain and upland landscape reflects a juxtaposition of landform and soil generations related to a shallow basin inversion, the last stages of which occurred in semi-arid conditions since the Oligocene. These results throw light on old debates on models of long-term landform development in platform areas, and also help in evaluating recent models of denudation and burial based on thermochronological methods. As a whole, this work brings new scientific data that further help in assessing geodiversity of a large territory, particularly on geomorphodiversity (landforms and processes) and pedodiversity aspects (soils and paleosoils). At a finer scale, it may also help in identifying valuable geomorphological sites (or geomorphosites) whose scientific and other values (cultural, aesthetic, ecological…) might justify their integration into the national network of Brazilian geosites. It also helps in identifying the distribution of hazards and risks related to landforms and geodynamics in northeast Brazil.

Acknowledgments

This book is the result of more than 15 years of fieldwork in the Ceará and adjoining states (Rio Grande do Norte, Paraíba, Pernambuco, Piauí), which were made possible mainly by financial and technical help of the authors' institutions (Universities of Paris-Sorbonne and Paris-Diderot; Laboratoire de Géographie Physique, UMR CNRS 8591; Laboratoire PRODIG, UMR CNRS 8586) and by the cooperation between the departments of Geography of the University of Paris-Sorbonne and of the Federal University of Ceará (UFC, Programa de Pós-Graduação, Fortaleza). Significant help was also provided by the Federal University of Rio Grande do Norte (UFRN, Natal), the Regional University of Cariri (URCA, Crato), and the Geopark Araripe (Crato). Among the numerous colleagues to whom the authors are indebted for their direct or indirect help, they wish to cite, in chronological order, the Professors and Doctors Vanda Claudino Sales, Jeovah Meireles, Michel Arthaud, Mariano Castelo Branco, Eustógio Dantas, Jean-René Vanney, Francisco Hilario Bezerra, Benjamin Bley de Brito Neves, Rubson Pinheiro Maia, Yanni Gunnell, Gérard Bourgeon, Laurent Caner, Alexsandra de Oliveira Magalhães, André Herzog, Francisco Idalécio Freitas, Rosiane Lima Verde, Francisco da Chagas Sousa da Costa, Ana Luiza Coelho Netto. They also wish to thank the numerous students and guides who participated in many fieldwork sessions. The initial versions of various figures were preparated with the help of Florence Bonnaud (University of Paris-Sorbonne). Finally, the authors warmly thank Jorge Rabassa, who, beyond offering them a participation in his research program on paleosurfaces in Argentina, invited them to present this manuscript in the Springer Earth System Sciences collection.

Contents

Abstract

More than a simple monograph, the authors present a comprehensive geomorphic overview of a large tropical region where they show how deciphering the long-term landform evolution helps in understanding the present set of landscapes and morphodynamic environments. The Equatorial margin of the Brazilian "Nordeste" displays stratigraphic landmarks whose interpretation reveals the age and nature of landforms, leading to a reconstruction of the geomorphic history by means of combined morphostratigraphic and morphopedological approaches. Beyond the role of differential erosion related to moderate post-oceanic opening uplift, the plain and upland landscape reflects a juxtaposition of landform and soil generations related to a shallow basin inversion, the last stages of which occurred in semi-arid conditions since the Oligocene. These results throw light on old debates on models of long-term landform development in platform areas, and also help evaluating recent models of denudation and burial based on thermochronological methods. As a whole, this work brings new scientific data that further help assessing geodiversity of a large territory, particularly on geomorphodiversity (landforms and processes) and pedodiversity aspects (soils and paleosoils). At a finer scale, it may also help identifying valuable geomorphological sites (or geomorphosites) whose scientific and other values (cultural, aesthetic, ecological…) might justify their integration into the national network of Brazilian geosites.

Keywords Landscape evolution · Morphostratigraphy · Morphopedology · Morphostructural patterns · Northeast Brazil · Passive margin · Geodiversity

Chapter 1
Introduction

Abstract The equatorial margin of the Brazilian "Nordeste" is especially rich in stratigraphic landmarks that help in understanding the nature of landforms and reconstructing the geomorphic history. It illustrates the interest of combined morphostructural, morphostratigraphic, and morphopedological approaches in the study of a large area ($\sim$200,000 km^2). The objectives are: (1) to propose a review of the geological data; (2) to analyze and revise the morphostructural and morphostratigraphic patterns; (3) to investigate features that might be specifically related to the complex history recorded by a passive margin intersecting a failed rift arm; (4) to show that this type of study is an irreplaceable tool for further geomorphological and geological researches on basement areas, passive margins, and related sedimentary basins; (5) to give a scientific substrate to an inventory of geodiversity in this region particularly rich in remarkable geomorphic and geological features.

Keywords Morphostructural analysis · Morphostratigraphy · Morphopedology · Passive margin · Geodiversity · Northeast Brazil

1.1 General Objectives

The equatorial margin of the Brazilian Nordeste differs from many uplifted passive margins often poor in Phanerozoic deposits and weathering formations, as it displays lots of stratigraphic landmarks that help in understanding the nature of landforms and reconstructing the geomorphic history. It provides a precious opportunity to illustrate the interest of combined morphostructural, morphostratigraphic, and morphopedological approaches in the study of a large area ($\sim$200,000 km^2).

Recent publications on the structure and morphology of the Brazilian continental margin mainly concern the eastern Atlantic front, a typical rifted margin with Great Escarpments (Ollier 1985a, b; Gallagher et al. 1994; Almeida and Carneiro 1998; Brown et al. 2000; Japsen et al. 2012; Jelinek et al. 2014). Although the structure of

J.-P. Peulvast and F. Bétard, *Landforms and Landscape Evolution of the Equatorial Margin of Northeast Brazil*, Springer Earth System Sciences,
DOI 10.1007/978-3-319-18203-2_1

equatorial northeastern Brazil and of conjugate parts of the African margin have also been analyzed in many recent papers (e.g., Popoff 1988; Magnavita et al. 1994; Matos 1992, 2000, 2002), the continental landforms of this region—a transform margin, i.e., resulting from wrench faulting and shearing between two continents before oceanic opening (Boillot and Coulon 1998; Mascle and Basile 1998)—have only partly been reinterpreted until now (Peulvast et al. 2008).

This margin segment strongly differs from those of eastern Brazil, forming a particular case among passive margins with rifted or "type 1" Great Escarpments (Beaumont et al. 2000; Pazzaglia and Gardner 2000). The presence of typical climatic landforms (pediplains, inselbergs…), stepped patterns and related paleolandforms makes this area a highly representative place for tropical, basement, and margin morphology.

More than a simple monograph, we present a comprehensive geomorphic overview of a large tropical region in order to show how deciphering the long-term landform evolution helps in understanding the present set of landscapes and morphodynamic environments. The Equatorial margin of the Brazilian "Nordeste" displays stratigraphic landmarks whose interpretation reveals the age and nature of landforms, leading to a reconstruction of the long-term geomorphic history by means of combined morphostratigraphic and morphopedological approaches. Beyond the role of differential erosion related to moderate post-oceanic opening uplift, the plain and upland landscape reflects a juxtaposition of landform and soil generations related to a shallow basin inversion, the last stages of which occurred in semi-arid conditions since the Oligocene. The results of this analysis are also expected to throw light on old debates on models of long-term landform development in platform areas, and help in evaluating recent models of denudation and burial based on thermochronological methods (Japsen et al. 2012; Jelinek et al. 2014).

Therefore, the book is organized around the following objectives: (1) to propose a review of the geological data that are necessary for understanding inland morphology; (2) to analyze morphostructural and morphostratigraphic patterns of this vast area whose only large morphological outlines are hitherto described; (3) to investigate features that might be specifically related to the complex history recorded by a passive margin intersecting a failed arm; (4) to show that, generally speaking, this type of study is an irreplaceable tool for further geomorphological and geological researches on basement areas, passive margins, and related sedimentary basins, for instance, by deciphering the respective influences of superposed structural heritages on the relief, discriminating tectonic, climatic, and other factors of geomorphological differentiation, giving bases for reconstructing the morphotectonic evolution or understanding the recent or current morphodynamic evolution; (5) to give a scientific substrate to an inventory of geodiversity which, in this region particularly rich in remarkable geomorphic and geological features, is still far from to be completed, in spite of the existence of various types of preservation units (Geopark Araripe, National Parks) and of the SIGEP preliminary inventory (Bétard et al. 2011).

1.2 Limits and Main Physiographic Features of the Study Area

The study area belongs to the Borborema Province (Almeida et al. 1981) and forms the northern part of the Brazilian northeast, or "Nordeste." This eastern segment of the Brazilian Equatorial margin is a passive margin whose counterpart, on the opposite side of the Equatorial Atlantic Ocean, is the northern shore of the Gulf of Guinea between Cameroon and Ghana (Popoff 1988; Maurin and Guiraud 1993; Matos 2000). Situated between 3°15′ and 7°30′S, and 37° and 41°W (560 × 520 km), it includes the Ceará state and the western parts of the Rio Grande do Norte and Paraiba states, as well as a northern fringe of the inner Pernambuco state, to the south (Fig. 1.1). Its western and southern limits, respectively, coincide with high plateaus, the Serra de Ibiapaba and the Chapada do Araripe (ca 1000 m a.s.l.). To the southeast and the east, it is bounded by the Borborema massif (1200 m a.s.l.), which separates it from the lower São Francisco valley and from the stepped plateaus and lowlands of the east coast, south of the São Roque Cape. The region belongs to the semi-arid Nordeste, but the coastal strip and the inner mountains are characterized by more humid tropical climates, with forest environments (Figueiredo 1997; Cavalcante 2005; Bétard 2007).

Along the ca 600 km long, WE to NW–SE coastline, the continental shelf is relatively narrow (from 100 km wide in the west to 41 km in the east: Freire et al. 2002) and shallow (50–80 m at the continental edge; Figs. 1.1 and 1.2). Most coastal landforms are low, with long tracts of sand beaches and barriers, dunes and narrow coastal plains locally interrupted by estuaries or by low rocky promontories and short active sea-cliffs (eastern Ceará), and often bounded inland by low abandoned cliffs (Claudino Sales 2002; Fig. 1.3a). Inland, a 10–80 km wide strip of low sedimentary plateaus called "tabuleiros" (10–60 m in altitude) is almost interrupted near Fortaleza, where coastal mountains rise up to 800 m, only a few kilometers far from the shoreline (Serra de Camara, Fig. 1.3b). Including a calcareous "chapada" east of the Jaguaribe River, these slightly dissected lowlands form the transition between the coast and the vast erosion plains—the so-called «Sertaneja Depression»—which extend far inland on the Precambrian basement, up to the foot of remote scarps and inselbergs.

The regional relief is organized around a large depression drained by the lower Jaguaribe river and its tributaries To the west, a wide highland zone drained by shorter rivers oriented to the north is characterized by summit altitudes around 1000 m a.s.l. To the north of the Borborema plateau, the eastern upland zone is narrow, with lower summits (700–900 m a.s.l.). At regional scale, the inland area, or "sertão," exhibits the roughly concentric pattern of a huge hemicycle (Fig. 1.4), here named "the Jaguaribe-Piranhas hemicycle," intersected by two discontinuous and offset EW alignments of mountain slopes forming a highly dissected escarpment

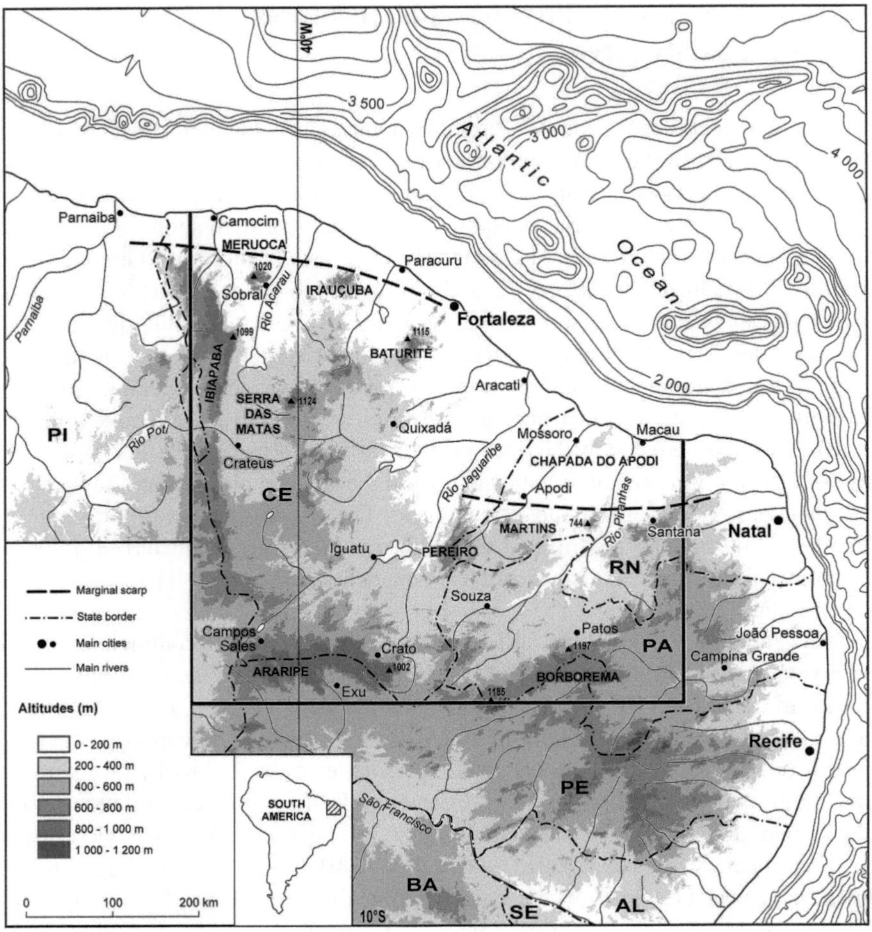

Fig. 1.1 Location map of the study area. Topography derived from the SRTM DEM (shuttle radar topography mission digital elevation model)

along the margin, perhaps a low and very discontinuous version of a Great Escarpment (in the sense of Ollier 1985a). The offset area corresponds to the lower Jaguaribe valley and plain, and opens on a SE-trending segment of the continental margin (Fortaleza - Potiguar). The highlands are arranged in two concentric half rings around the central plain which widens toward the Aracati Platform and the Potiguar Basin (Fig. 1.4). Historically, less attention has been paid to this geographical pattern than to the stepped patterns described or postulated by many authors (Peulvast and Claudino Sales 2004).

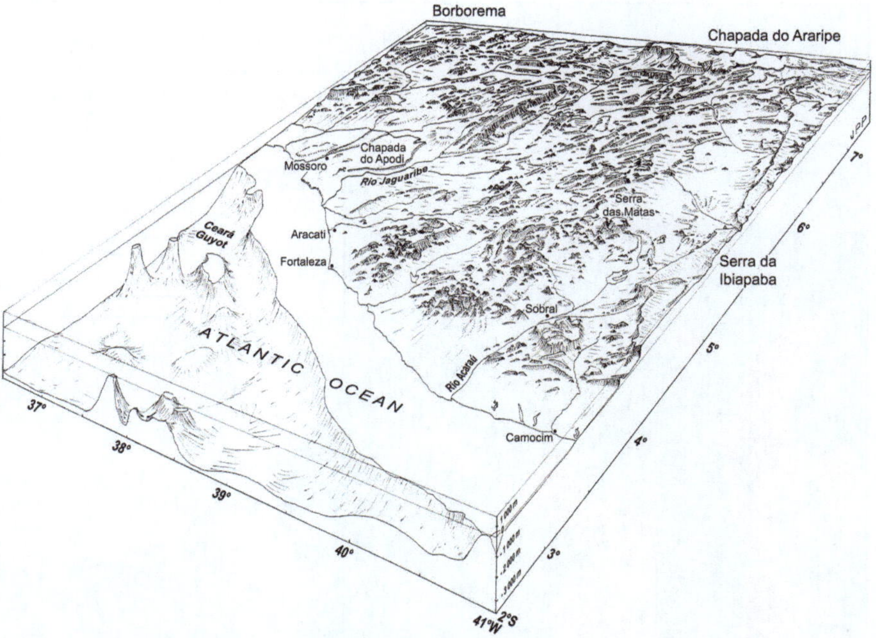

Fig. 1.2 Sketch block-diagram of the Jaguaribe—Piranhas amphitheater, showing the main topographic features

Fig. 1.3 **a** Guriú, west of Jericoacoara: wide tabuleiros interrupted over the sea by a low abandoned sea-cliff (*left*) widely overriden by mobile dunes (barchanoids). **b** Icaraí de Caucaia (western outskirts of Fortaleza): narrow coastal plain fringed by sandy shore and dunes, and overlooked to the south by residual mountains exceptionally close to the coast (Serra de Juá, Serra da Conceição). *Photographs* J.P. Peulvast

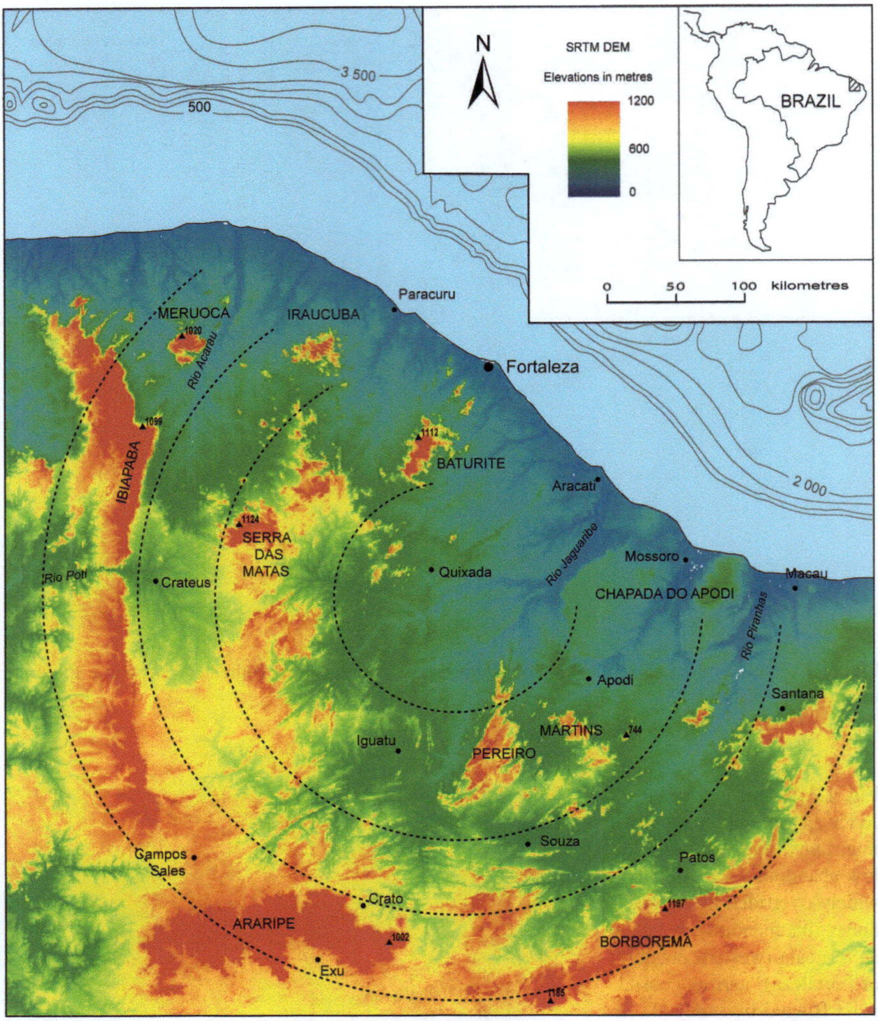

Fig. 1.4 Topography of the study area. Note the concentric patterns of landforms, organized in two half rings of highlands around a wide central plain—the so-called "Sertaneja Depression." Topography derived from the SRTM DEM

1.3 The Regional Organization of Landforms: Classical Interpretations

1.3.1 Morphological Units and Previous Classifications

In the literature, the morphological features of the study area are either described on the basis of stratigraphic and topographic criterions or in relation to tectonics. One of the most detailed descriptions and partitioning of the regional morphology was proposed in the Radambrasil Project (Moreira and Gatto 1981; Prattes et al. 1981). Here, these authors identify the following morphostructural units:

1. the coastal plain,
2. the coastal plateaus or "tabuleiros" underlain by the soft sandstones of the Barreiras Formation,
3. the low Sertaneja Surface, widely developed inland around 350 m a.s.l. and lower, and considered to be the fundamental topographic level,
4. the residual massifs or plateaus, above 700 m a.s.l., which correspond to the mountains of the inner half ring of highlands, and to the marginal escarpment,
5. the Ibiapaba Plateau.

Strong structural controls appear in parts of the outlines of these units and in the dissected landforms.

Another classification of morphological domains, proposed for the Ceará state (Souza 1988, 1997), reflects the underlying lithological and chronostratigraphic units:

- the domain of Cenozoic deposits (fluvial plains, coastal landforms, tabuleiros or "Glacis pré-litorâneos"),
- the domain of Paleo-Mesozoic sedimentary basins, with related structural landforms (Chapada do Araripe, Chapada do Apodi, Ibiapaba plateau),
- the basement units, comprising the dissected residual mountains and the Sertaneja depressions, slightly dissected and strewn with inselbergs.

The classification proposed by Saadi and Torquato (1992) also concerns the Ceará state. It opposes "peripheric" sedimentary basins (in fact at the periphery of the state) and a "Central crystalline block or nucleus" (the western part of the inner mountain half ring). The Sertaneja surface and the "residual mountains" would register a regional updoming north of the EW Patos shear zone, between NE-trending depressions or grabens located in northwest (Jaibaras) and southeast Ceará (Senador Pompeu, Orós and Jaguaribe fault systems). These descriptions do not account for the roughly concentric patterns described above.

1.3.2 Stepped Surfaces: Classical Models of Interpretation

The morphological patterns of northeast Brazil are classically considered as a result of regional uplift induced by break-up of the Gondwana supercontinent, interfering with climatic fluctuations in the Cenozoic (Mabesoone and Castro 1975). The uplift would have triggered the shaping of successively lower planation surfaces and the incision of a centrifugal drainage pattern (Fig. 1.5).

As in the Guiana Shield (McConnell 1968; Zonneveld 1985, 1993), all authors agree on the existence of stepped erosion surfaces in this region, related to Mesozoic and Cenozoic sediments deposited on the continental margin. Landscape patterns have been interpreted as (i) the erosional response to updoming of a large crustal

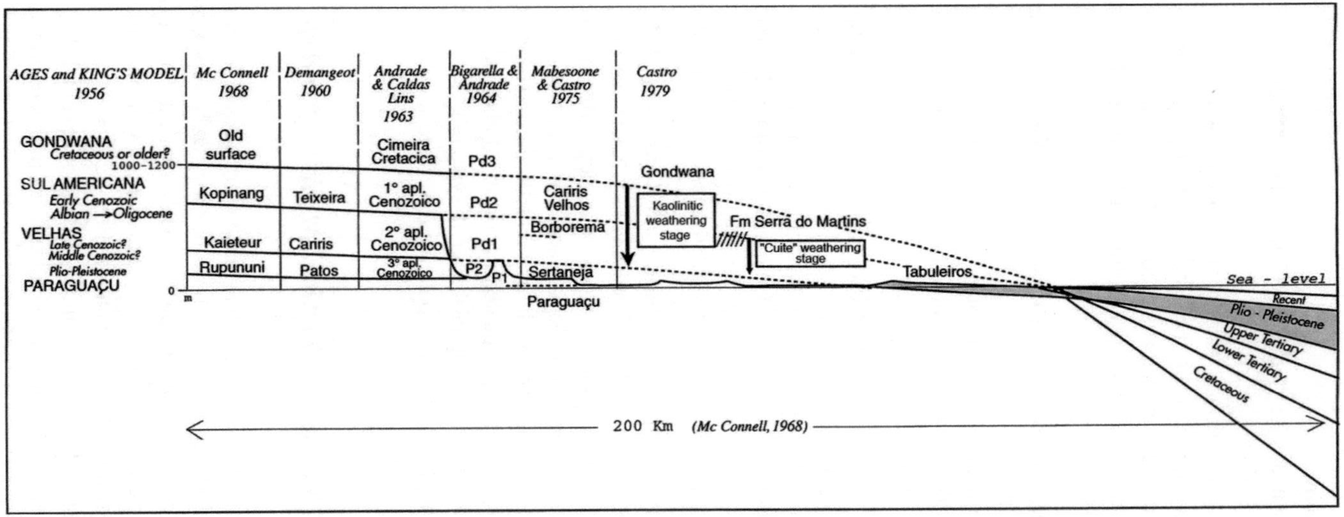

Fig. 1.5 Patterns and correlation of planation surfaces in northeastern Brazil and adjoining regions

swell (Ab'Saber 1956; Dresch 1957; Demangeot 1960; Andrade and Caldas Lins 1965), (ii) differential uplift (Bigarella and Andrade 1964) and/or (iii) a seaward 'continental flexure' (Ruellan 1952; King 1956). Regional uplift would have shaped a descending staircase of successively younger erosion surfaces driven by an incising radial drainage pattern and oscillations between wetter and drier climates during the Cenozoic. As on the nearby shield regions of Guiana (McConnell 1968; Zonneveld 1985, 1993), most authors have recognized the treads of four surfaces in this region, and have linked them to Mesozoic and Cenozoic offshore and onshore sedimentary sequences (Fig. 1.5) and to epeirogenic uplift with a crest located 150–200 km inland. According to Ab'Sáber (2000), the Borborema plateau, surrounded and cut by the depressions of the Sertaneja surface, and considered as the central massif of northeast Brazil, would correspond to a "shield dome" and its westward dipping flank. However, Demangeot (1960) had suggested that Cretaceous and later faulting and differential uplift might have occurred, warping the sub-Cretaceous surface between 250 and 1000 m in inner parts of the Pernambuco, Paraiba and Ceará states.

Many reconstructions of paleosurfaces were elaborated in eastern and southeast Brazil (Passos and Bigarella 1998; Ab'Saber 2000) and merely adapted to the Nordeste with local names (Teixeira, Borborema, Patos, Sertaneja surfaces). It is often difficult to find a good fit between these theoretical models and the real landforms in the field. Moreover, geological profiles are often missing, and geomorphological maps are generally inadequate for such a study, since no attempt is made to map the surfaces (except for Demangeot 1960, and the RadamBrasil geomorphological map: Moreira and Gatto 1981; Prattes et al. 1981). At last, obviously exhumed paleosurfaces have never been taken into account in any reconstruction—e.g., the exhumed pediplain and inselbergs exposed below the western cuesta of the Chapada do Apodi (Maio 1987).

1.4 The Morphostructural, Morphostratigraphic, and Morphopedological Approaches

1.4.1 Morphostructural Patterns, Morphostratigraphy and Landform Ages

Our approach is based on geomorphological mapping, and involves a systematic interpretation of morphostructural patterns (i.e., a definition of landforms according to their relationships with tectonic and lithological factors at various scales) and of morphostratigraphic relationships (e.g., burying and exhumation of paleolandforms and unconformities). Data were collected from topographic and geological maps at various scales, including the geological map of Ceará at 1:500,000 (CPRM 2003) and the geological map of Brazil at 1:2,500,000 (CPRM 2001). We also used the SRTM DEM (Shuttle Radar Topography Mission Digital Elevation Model), combined with geological and other data in a Geographic Information System (GIS).

Extensive field studies were completed by interpretation of air photographs, airborne radar, and Landsat images, and by drawing of geological profiles and block-diagrams. Combined with drainage anomalies analysis, this approach leads us to define an original interpretation of the regional landforms and of the marginal escarpment, and to discuss the nature and significance of the tectonic and lithological controls on morphological patterns.

By analogy with stratigraphy, morphostratigraphy can be defined as the study of the succession of landforms (Peulvast et al. 2009). It gives relative chronology by adapting the principles of continuity (a given landform is supposed to present the same age in all its parts) and superposition, or substitution (a given landform is older than the landforms that tend to obliterate it). The underlying concept is that landforms are time-dependent and have an age. In fact such an assertion is not true in all cases, since many landforms are constantly reworked and time-independent, either in the "acyclic" cratonic areas (Klein 1997) or in mountain ranges evolving in steady-state conditions.

The idea that the "age" of a landform or landsurface implies minimal change by denudation processes since its creation (Summerfield et al. 1999). In platform areas as well as in other structural contexts, the most precise ages are obtained on original structural landforms (Peulvast and Vanney 2001). Corresponding to specific geological dynamics, they were formed during very short events (earthquake, volcanic eruption, meteorite impact) or sequences of processes (e.g., faulting or folding) that can be dated by relative or absolute methods. The original landforms have the same age as the corresponding structure or the ultimate stage of its formation. In favorable conditions, the age of the simplest of these landforms (co-seismic scarps…) may be deduced from their state of preservation.

The age of erosional landforms is less well-defined. A priori, it corresponds to the period when they ceased to evolve. However, the age of a given landform can differ from one part to another (e.g., between the distal and proximal parts of a pediment). Among other conditions, some structural landforms may be created by differential erosion after dissection of a planation surface or through exhumation of buried landscapes. Their age is well known when they relate to a well dated generation of cyclic landforms or a climatic sequence (e.g., the structural landforms of glacial origin identified in east Greenland: Peulvast and Vanney 2001). Therefore, inherited or fossil structural landforms (e.g., the paleo-hogbacks of southern Gaspésie, Quebec, which are still partly sealed by Visean conglomerates: Peulvast et al. 1996) give the best contribution to landform dating and geomorphic reconstructions.

Some morphogenetic events which create new landforms may be easily identified thanks to stratigraphic markers with a well-known geometry and age. Primitive structural surfaces or aggradation surfaces are also useful, if well dated. They may be deformed without being immediately eroded, especially if the deformations correspond to faulting or vertical movements. Analyzing these deformations and their geomorphic effects is more difficult if stratigraphic markers are lacking, as frequently occurs in basement areas: other landmarks must be used (Summerfield 1987). Initially flat surfaces, especially planation surfaces are considered to be the most convenient landmarks to reconstruct the local or regional tectonic and geomorphic history, provided that they are well identified, related with a former base level and dated (Calvet and Gunnell 2008; Japsen et al. 2009). When well accomplished, planation processes have roughly the same obliterating effects on older topographies as aggradation processes. However, the age of planation surfaces is more difficult to obtain, except if unconformable deposits seal these surfaces. Although some authors assume that planation surfaces may be simultaneously developed at different altitudes in the same region in relation with separate base levels (African Surface of southern Africa: Partridge and Maud 1987), stepped systems of undeformed surfaces are generally considered as presenting a chronological meaning (Lageat and Robb 1984; Klein 1997). Elements of various ages and origins often coexist in apparently uniform topographies (Peulvast and Claudino Sales 2005). Uncertainties remain about their age, initial, terminal, before burying or dissection, or since exhumation. They often are diachronic, particularly if they result from scarp retreat. The residuals and scarps that bound them may correspond to lithological controls, without chronological meaning. Some surfaces cannot be dated since they are constantly reworked in the conditions of acyclic regime or slow degradation (Klein 1997).

1.4.2 Morphopedology and Duricrusts: Chronological Implications

The morphopedological approach was used and improved by one of us (Bétard 2007) in his study of the Baturité massif (northern Ceará), and more recently in a study of the history of the topographic inversion of the Araripe basin, in southern Ceará (Peulvast and Bétard 2015). Its objective is to show the complex interrelations between landforms and soils, through an integrated study which combines both geomorphological and pedological investigations. Initially employed as a tool for applied research in soil science for producing soil maps (Bétard and Bourgeon 2009), this approach can be used today as a scientific method to improve a regional geomorphological study, particularly in basement regions where soils and weathering formations are often the only indicators and recorders of long-term landscape evolution.

Observational data on soils and weathered materials were collected in the field and from existing soil maps and previous inventories (Guichard 1970; Projeto Radambrasil: Moreira and Gatto 1981; Prattes et al. 1981). Special attention was paid to the distribution of laterite (i.e., duricrusts), which is commonly considered as a reliable marker in the reconstruction of denudation histories and can be correlated with regional paleoclimates (Tardy and Roquin 1998). Laterite mapping at regional scale was completed using a combination of field surveys and the processing of Landsat imagery and digital elevation models (SRTM DEM). Analyzing the relationship between stepped surfaces, laterite distribution, and soil development provides possible correlations between the maturity of soil or weathering profiles and the different generations of landforms in the multi-storeyed landscape.

Soils and characteristic weathering formations (saprolites, ferricretes, calcretes...) may help dating planation surfaces or differentiated landscapes (Godard et al. 2001; Vasconcelos and Conroy 2003; Watchman and Twidale 2002). However, such dating is delicate because superficial formations are not always correlative of the planation process, which can be older. Moreover formations, which could be dated by radiochronological methods, as the extensive laterite covers in tropical regions, undergo a continuous geochemical evolution. They cannot be considered as closed systems and stable markers (Gunnell 2003; Nahon 2003).

According to geochemical calculations, the long durations implied in the transformation of kaolinized weathering horizons into ferricretes lead to consider old ages for the underlying topographies. For example, the "haut-glacis" and "moyen-glacis" of western Africa, respectively considered as Early and mid Quaternary in age in former French publications, are older than 8 and 4 Myear (Tardy and Roquin 1998). Indeed, dating lateritic crusts by the means of the paleomagnetic method and by oxygen isotopes generally confirms their 'old' ages. In West Africa just as in Brazil, the laterites observed in the plane of the African or Sul-Americana surface, considered as Eocene in age, were effectively dated at ~50 Myear (Hénocque et al. 1998; Tardy and Roquin 1998; Gunnell 2003). In eastern Ceará (Pereiro plateau), they are possibly as old as 20 Myear (Lima 2008). However, Although they are considered by authors who studied this area as markers of vertical movements, assuming that they should be formed on surfaces related to the base level, they do not clearly support evidence of a strong local uplift and cannot be considered as good markers of vertical tectonics, since laterite can form in various topographic and altitudinal contexts (Rossetti 2004).

1.5 A Contribution to the Knowledge of Geodiversity in Northeast Brazil

The concept of geodiversity is quite recent and can be defined as "the natural range (diversity) of geological (rocks, minerals, fossils), geomorphological (landforms, processes) and soil features. It includes their assemblages, relationships, properties, interpretations and systems" (Gray 2004). Whereas geodiversity refers to the variety of abiotic nature, geoheritage is the set of the most relevant geodiversity elements with particular importance for science, education, or tourism (Pereira et al. 2012). During recent years, some attempts were made in order to map geodiversity and to inventory geoheritage and geosites in the territories at various scales. For instance in Brazil, the National Geological Service (CPRM) published the Geodiversity Map of Brazil at 1:2,500,000 scale (CPRM 2006), which is a synthesis of the major geosystems that constitute the national territory, as well as their limitations and potential uses for geotourism or georessources exploitation (Silva 2008). However, the CPRM geodiversity map is only based on geological data such as lithostratigraphy and mineral resources and does not take into account other geodiversity elements such as landforms and soils, which are also important to interpret the whole geodiversity and to support decision making and land-use management issues (Pereira et al. 2012).

Our study is intended to give new scientific data that help in integrating other elements of geodiversity than the geological diversity *stricto* sensu, both on geomorphodiversity (morphostructural patters, climatic landforms, and paleolandforms, morphodynamic features) and pedodiversity aspects (soils and paleosoils, laterites and duricrusts). At a finer scale, it may also help in identifying valuable geomorphological sites (or geomorphosites) whose scientific and other values might justify their integration into the national network of Brazilian geosites (SIGEP inventory: http://sigep.cprm.gov.br/).

References

Ab'Saber AN (1956) Depressões periféricas e depressões semi-áridas do Nordeste do Brasil. Bol Paulista de Geogr 22:3–18
Ab'Saber AN (2000) Summit surfaces in Brazil. Rev Bras Geociências 30(3):515–516
Almeida FFM, Carneiro CDR (1998) Origem e evolução da Serra do Mar. Rev Bras Geociências 28(2):135–150
Almeida FFM, Hasui Y, Brito Neves BB, Fuck RA (1981) Brazilian structural provinces: an introduction. Earth Sci Rev 17:1–21
Andrade GO, Caldas Lins R (1965) Introdução à morphoclimatologia do Nordeste do Brasil (2a ed., revista). Inst Ciên Terra, Univ. Recife 3–4:17–28
Beaumont C, Kooi H, Willett S (2000) Coupled tectonic-surface process models with applications to rifted margins and collisional orogens. In: Summerfield MA (ed) Geomorphology and global tectonics, Wiley, Chichester, pp 29–55
Bétard F (2007) Montagnes humides au cœur du Nordeste brésilien semi-aride: le cas du massif de Baturité (Ceará). Apports d'une approche morphopédologique pour la connaissance et la gestion des milieux. Ph.D. thesis, Univ Paris-Sorbonne
Bétard F, Bourgeon G (2009) Cartographie morphopédologique: de l'évaluation des terres à la recherche en géomorphologie. Géomorphologie: relief, processus, environnement 3:187–198
Bétard F, Peulvast JP, Magalhães AO (2011) Biodiversité, géodiversité et enjeux de leur conservation dans les montagnes humides du Nordeste brésilien. Bull Assoc Géogr Fr 88(1):17–26
Bigarella JJ, Andrade GO (1964) Considerações sobre a estratigrafia dos sedimentos cenozóicos em Pernambuco (grupo Barreiras). Univ. Recife, Inst. Ciên. Terra, Arquivos 2:2–14
Boillot G, Coulon C (1998) La déchirure continentale et l'ouverture océanique. Géologie des marges passives. Gordon and Breach Sci Publ, Amsterdam
Brown RW, Gallagher K, Gleadow AJW, Summerfield MA (2000) Morphotectonic evolution of the South Atlantic margins of Africa and South America. In: Summerfield MA (ed) Geomorphology and global tectonics. Wiley, Chichester, pp 255–281
Calvet M, Gunnell Y (2008) Planar landforms as markers of denudation chronology: an inversion of East Pyrenean tectonics based on landscape and sedimentary basin analysis. In: Gallagher K, Jones SJ, Wainwright J (eds) Landscape evolution: denudation, climate and tectonics over different time and space scales. Geological Society, London, Special Publications 296, pp 147–166
Cavalcante A (2005) Jardins suspensos no Sertão. Sci Am Bras 32:69–73

Claudino Sales V (2002) Evolution géomorphologique de la zone côtière de Ceará (Nord-est du Brésil): du long terme au court terme. Ph.D. thesis, Univ Paris-Sorbonne
CPRM-Serviço Geológico do Brasil (2001) Geologia, tectônica e recursos minerais do Brasil: Sistema de Informações Geográficas – SIG e mapas na escala 1:2,500,000. Serviço Geológico do Brasil, 4 CD Rom
CPRM-Serviço Geológico do Brasil (2003) Atlas digital de geologia e recursos minerais do Ceará. Mapas na escala 1:500,000. Serviço Geológico do Brasil, CD Rom
CPRM-Serviço Geológico do Brasil (2006) Mapa geodiversidade do Brasil. CPRM, Brasília
Demangeot J (1960) Essai sur le relief du Nord-Est brésilien. Ann Géogr 69(372):157–176
Dresch J (1957) Les problèmes morphologiques du Nord-Est brésilien. Z Geomorph 1:289–301
Figueiredo MA (1997) Unidades fitoecológicas. In: IPLANCE (ed) Atlas do Ceará, Fortaleza, pp 28–29
Freire GSS, Filho WFS, Gomes DF, Lima SF, Monteiro LHU (2002) Provincias sedimentares da plataforma continental do Nordeste do Brasil. XLI Congr Bras Geol, João Pessoa, Anais, p 90
Gallagher K, HawkesWorth CJ, Mantovani MSM (1994) The denudation history of the onshore continental margin of SE Brazil inferred from apatite fission track data. J Geophys Res 99:18117–18145
Godard A, Lagasquie JJ, Lageat Y (2001) Basement regions. Springer, Berlin
Gray M (2004) Geodiversity: valuing and conserving abiotic nature. Wiley, Chichester
Guichard E (1970) Les sols du Bassin du Rio Jaguaribe. Mémoire ORSTOM 40, Paris
Gunnell Y (2003) Radiometric ages of laterites and constraints on long-term denudation rates in West Africa. Geology 31:131–134
Hénocque O, Ruffet G, Colin F, Féraud G (1998) $^{40}Ar/^{39}Ar$ dating of West African lateritic cryptomelanes. Geochim et Cosmochim Acta 62:2739–2756
Japsen P, Bonow J, Green P, Chalmers JA, Lidmar-Bergström K (2009) Formation, uplift and dissection of planation surfaces at passive continental margins—a new approach. Earth Surf Proc Land 34(5):683–699. doi:10.1002/esp.1766
Japsen P, Bonow JM, Green PF, Cobbold PR, Chiossi D, Lilletveit R, Magnavita LP, Pedreira A (2012) Episodic burial and exhumation in NE Brazil after opening of the South Atlantic. Geol Soc Am Bull 124:800–816
Jelinek AR, Chemale F Jr, van der Beek PA, Guadagnin F, Cupertino JA, Viana A (2014) Denudation history and landscape evolution of the Northern East-Brazilian continental margin from apatite fission-track thermochronology. J S Am Earth Sci 54:158–181
King LC (1956) A geomorfologia do Brasil oriental. Rev Bras Geografia 18(2):3–121
Klein C (1997) Du polycyclisme à l'acyclisme en, géomorphologie edn. Ophrys, Gap
Lageat Y, Robb LJ (1984) The relationships between structural landforms, erosion surfaces and the geology of the Archean granite basement in the Barberton region, Eastern Transvaal. Trans Geol Soc S Afr 87:141–159
Lima MG (2008) A História do Intemperismo na Província Borborema Oriental, Nordeste do Brasil. Implicações paleoclimáticas e tectônicas. Ph.D. thesis, Universidade Federal do Rio Grande do Norte, Natal
Mabesoone JM, Castro C (1975) Desenvolvimento geomorfológico do nordeste brasileiro. Bol Núcleo Nordeste da SGB, Recife 3:5–35
Magnavita LP, Davison I, Kuznir NJ (1994) Rifting, erosion and uplift history of the Reconcavo-Tucano-Jatoba Rift, Northeast Brazil. Tectonics 13:367–388
Maio CR (1987) Geomorfologia do Brasil: fotos e comentários. IBGE, Rio de Janeiro
Mascle J, Basile C (1998) Marges continentales transformantes. CR Acad Sci Paris 326:827–838
Matos RMD (1992) The Northeast Brazilian rift system. Tectonics 11(4):766–791
Matos RMD (2000) Tectonic evolution of the Equatorial South Atlantic. In: Mohriak W, Taiwani M (eds) Atlantic Rifts and continental margins. Geophys Monograph 115, Am Geophys Union, pp 331–354
Matos RMD (2002) Geologia regional da margem equatorial afro-brasileira: evolução cinemática e o contexto paleogeográfico. XLI Congr Bras Geol, João Pessoa, Anais, p 272

Maurin JC, Guiraud R (1993) Basement control in the development of the early Cretaceous West and Central African rift system. Tectonophysics 228:81–95

McConnell RB (1968) Planation surfaces in Guyana. Geogr J 134(4):506–520

Moreira MMM, Gatto LCS (1981) Geomorfologia. Projeto RadamBrasil, Folha Fortaleza, DNPM/ MME, Brasilia 21:213–252

Nahon P (2003) Altérations dans la zone tropicale. Signification à travers les mécanismes anciens et/ou encore actuels. CR Geoscience 335:1109–1119

Ollier CD (1985a) Morphotectonics of passive continental margins: introduction. Z Geomorph NF Suppl Bd 54:1–9

Ollier CD (1985b) Morphotectonics of continental margins with great escarpments. In: Morisawa M, Hack JT (eds) Geomorphology and tectonics. Binghamton Symposium, pp 3–25

Partridge TC, Maud RR (1987) Geomorphic evolution of Southern Africa since the Mesozoic. S Afr Geol J 90(2):179–208

Passos E, Bigarella JJ (1998) Superfícies de erosão. In: da Cunha SB, Guerra AJT (eds) Geomorfologia do Brasil. Bertrand Brasil, Rio de Janeiro, pp 107–141

Pazzaglia FJ, Gardner TW (2000) Late Cenozoic landscape evolution of the US Atlantic passive margin: insights into a North American Great Escarpment. In: Summerfield MA (ed) Geomorphology and global tectonics, Wiley, Chichester, pp 283–301

Pereira DI, Santos L, Silva J, Pereira P, Brilha JB, Silva J, Rodrigues C (2012) Mapping regional geodiversity in Brazil and Portugal. Geologia dell'Ambiente, Suppl 3:107–108

Peulvast JP, Bétard F (2015) A history of basin inversion, scarp retreat and shallow denudation: the Araripe Basin as a keystone for understanding long-term landscape evolution in NE Brazil. Geomorphology 233:20–40. doi:10.1016/j.geomorph.2014.10.009

Peulvast JP, Bétard F, Lageat Y (2009) Long-term landscape evolution and denudation rates in shield and platform areas: a morphostratigraphic approach. Géomorphologie: relief, processus, environment 2:95–108. doi:10.4000/geomorphologie.7540

Peulvast JP, Bouchard M, Jolicoeur S, Pierre G, Schroeder J (1996) Palaeotopographies and post-orogenic morphotectonic evolution around the Baie des Chaleurs (Eastern Canada). Geomorphology 16:5–32

Peulvast JP, Claudino Sales V (2004) Stepped surfaces and palaeolandforms in the northern Brazilian "Nordeste": constraints on models of morphotectonic evolution. Geomorphology 62:89–122

Peulvast JP, Claudino Sales V (2005) Surfaces d'aplanissement et géodynamique. Géomorphologie: relief, processus, environnement 4:249–274

Peulvast JP, Claudino Sales V, Bétard F, Gunnell Y (2008) Low post-Cenomanian denudation depths across the Brazilian Northeast: implications for long-term landscape evolution àt a transform continental margin. Global and Planetary Change 62:39–60. doi:10.1016/j.gloplacha.2007.11.005

Peulvast JP, Vanney JR (2001) Géomorphologie structurale, tome 1. Relief et structure. Gordon & Breach, Paris, and BRGM Editions, Orléans

Popoff M (1988) Du Gondwana à l'Atlantique sud: les connexions du fossé de la Bénoué avec les bassins du NE brésilien jusqu'à l'ouverture du Golfe de Guinée au Crétacé inférieur. J Afr Earth Sci Spec Publ 7(2):409–431

Prattes M, Gatto LCS, Costa MIC (1981) Geomorfologia. In: Projeto RadamBrasil, vol 23, pp 301–348. Folha Jaguaribe/Natal, MME/DNPM, Brasilia

Rossetti DF (2004) Paleosurfaces from Northeastern Amazonia as a key for reconstructing paleolandscapes and understanding weathering products. Sed Geol 169:51–174

Ruellan F (1952) O escudo brasileiro e os dobramentos de fundo. Univ do Brasil, Rio de Janeiro, Depto Geogr Fac Nac Filosofia

Saadi A, Torquato JR (1992) Contribuição à neotectônica do Estado do Ceará. Rev Geologia 5:5–38

Silva CR (2008) Geodiversidade do Brasil: conhecer o passado, para entender o presente e prever o futuro. CPRM, Serviço Geológico do Brasil

Souza MJN (1988) Contribuição ao estudo das unidades morfo-estruturais do estado do Ceará. Rev Geologia 1(1):73–91

Souza MJN (1997) Geomorfologia. In: IPLANCE (ed) Atlas do Ceará, Fortaleza, pp 18–19
Summerfield MA (1987) Neotectonics and landform genesis. Prog Phys Geogr 11(3):384–397
Summerfield MA, Sugden DE, Denton GH, Marchant DR, Cockburn HAP, Stuart FM (1999) Cosmogenic isotope data support previous evidence of extremely low rates of denudation in the dry valleys region, Southern Victoria Land, Antarctica. Geol Soc Spec Pub 162:255–267
Tardy Y, Roquin C (1998) Dérive des continents, paléoclimats et altérations tropicales. Ed BRGM, Orléans
Vasconcelos PM, Conroy M (2003) Geochronology of weathering and landscape evolution, Dugald River valley, NW Queensland, Australia. Geochim et Cosmochim Acta 67:2913–2930
Watchman AL, Twidale CR (2002) Relative and "absolute" dating of land surfaces. Earth-Sci Rev 58:1–49. doi:10.1016/S0012-8252(01)00080-0
Zonneveld JIS (1985) Geomorphological notes on the continental border in the Guyanas (N. South America). Z Geomorph NF, Suppl Bd 54:71–83
Zonneveld JIS (1993) Planations and summit levels in Suriname (S. America). Z Geomorph NF, Suppl Bd 93:29–46

Chapter 2
Geological Setting: The Borborema Province

Abstract The study area is part of the Brazilian Equatorial continental margin, formed after the Equatorial Atlantic opening in Aptian times. Onshore, a Precambrian basement—the Borborema structural province—is subdivided into several geological domains by a wide system of Late Proterozoic shear zones partly reactivated in the Cretaceous. The present structural pattern is organized around a discontinuous NE–SW set of basins and half grabens—the intracratonic Cariri-Potiguar rift zone. Representing failed rift structures of Early Cretaceous age with remnants of their post-rift sedimentary cover, it is intersected by the transform margin in the Potiguar Basin area. Remains of the pre-rift cover of the basement are preserved outside these structures (the Parnaiba Basin, to the west) and inside the rift zone in southern Ceará (Araripe Basin). Along the coast, a thin dissected layer of Cenozoic continental sediments (the Barreiras Group) underlies the low-lying "tabuleiros" between the marginal escarpment and the shallow continental shelf.

Keywords Precambrian basement · Rifting · Oceanic opening · Transform margin · Clastic sedimentation · Volcanism · Northeast Brazil

The area described in this book is part of the Brazilian Equatorial continental margin, formed after the Equatorial Atlantic opening in Aptian times (Matos 2000). Onshore, a Precambrian basement—the Borborema structural province (Almeida et al. 1981; Caby et al. 1995; Arthaud et al. 2008)—is subdivided into several geological domains by a wide system of Late Proterozoic shear zones partly reactivated in the Cretaceous (Fig. 2.1). The present structural pattern is organized around a discontinuous NE–SW set of basins and half grabens—the 500-km long intracratonic Cariri-Potiguar rift zone. Representing failed rift structures of Early Cretaceous age with remnants of their post-rift sedimentary cover (Araripe and Potiguar basins, at both extremities), it is intersected by the transform margin in the Potiguar Basin area (Matos 1992). Remains of the pre-rift cover of the basement are preserved outside these structures (the Parnaiba Basin, to the west) and inside the rift zone in southern Ceará (Araripe Basin). Along the coast, a thin dissected layer of Cenozoic continental sediments (the Barreiras Group) underlies the low-lying "tabuleiros" between the marginal escarpment and the shallow continental shelf.

J.-P. Peulvast and F. Bétard, *Landforms and Landscape Evolution of the Equatorial Margin of Northeast Brazil*, Springer Earth System Sciences,
DOI 10.1007/978-3-319-18203-2_2

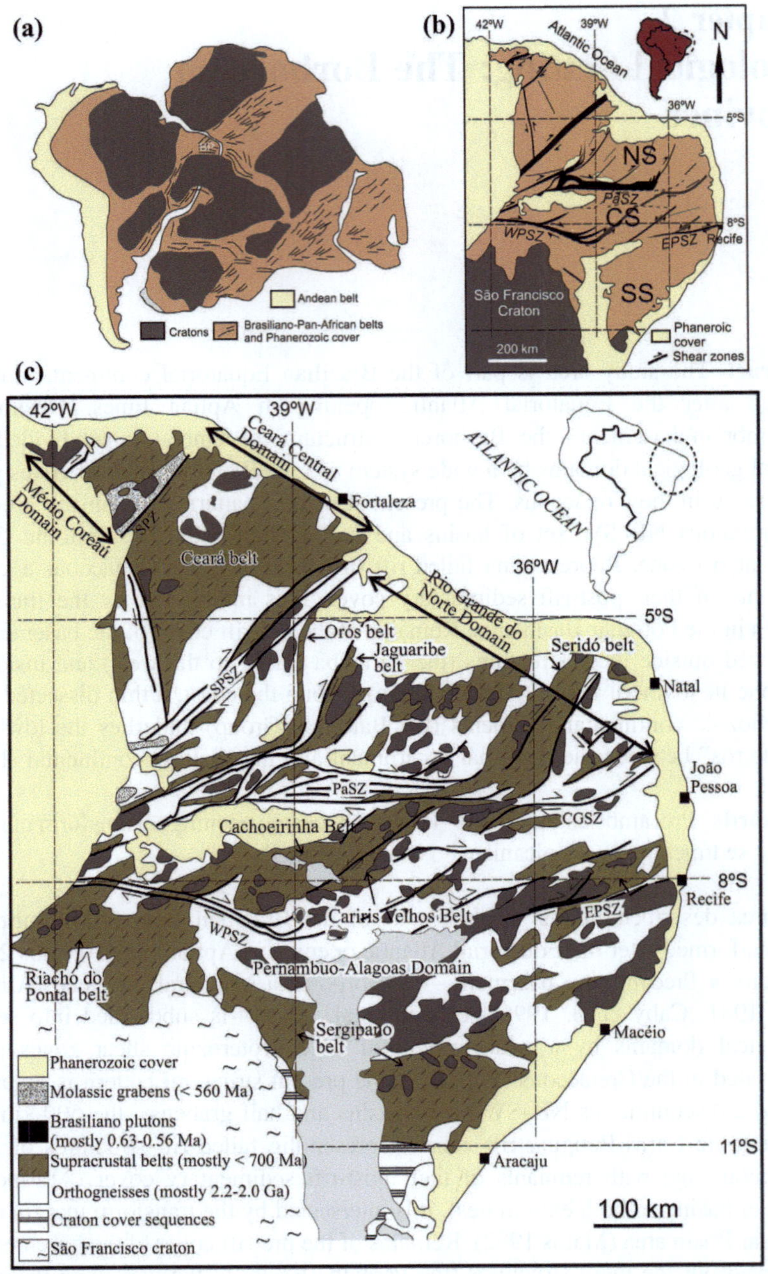
(a)
Andean belt
Cratons
Brasiliano-Pan-African belts and Phanerozoic cover
(b)
42ºW
39ºW
36ºW
Atlantic Ocean
N
5ºS
8ºS
NS
PaSZ
CS
WPSZ
EPSZ
Recife
SS
São Francisco Craton
200 km
Phaneroic cover
Shear zones
(c)
42ºW
39ºW
36ºW
5ºS
8ºS
11ºS
ATLANTIC OCEAN
Médio Coreaú Domain
Ceará Central Domain
Rio Grande do Norte Domain
Fortaleza
SPZ
Ceará belt
Orós belt
Jaguaribe belt
Seridó belt
Natal
SPSZ
João Pessoa
PaSZ
CGSZ
Cachoeirinha Belt
Recife
EPSZ
WPSZ
Cariris Velhos Belt
Pernambuo-Alagoas Domain
Riacho do Pontal belt
Maceió
Sergipano belt
Aracaju
100 km
Phanerozoic cover
Molassic grabens (< 560 Ma)
Brasiliano plutons (mostly 0.63-0.56 Ma)
Supracrustal belts (mostly < 700 Ma)
Orthogneisses (mostly 2.2-2.0 Ga)
Craton cover sequences
São Francisco craton

◀ **Fig. 2.1** **a** Pre-drift reconstruction of South America–Africa showing the Andean belt, Archean/Proterozoic cratons, and Brasiliano/Pan-African provinces of western Gondwana. *BP* Borborema Province. **b** Sketch showing the division of the Borborema Province in Northern (NS), Central (CS), and Southern (SS) subprovinces. Shear zone systems: PaSZ, Patos; EPSZ, East Pernambuco; WPSZ, West Pernambuco. **c** Schematic map of the Borborema Province showing the subdivision of the Northern and Southern subprovinces in the Medio Coreaú, Ceará Central and Rio Grande do Norte domains, and in the Pernambuco-Alagoas Domain and Sergipano and Riacho do Pontal belts, respectively. Main shear zones and supracrustal belts are named. Abbreviations for transcurrent shear zones: CGSZ, Campina Grande; EPSZ, East Pernambuco; PaSZ, Patos; SPSZ, Senador Pompeu; SPZ, Sobral; WPSZ, West Pernambuco. From Neves (2015)

Only a few elements of the morphology—fault scarps, structural surfaces—directly reflect this structural pattern. Most other morphostructural features, including the marginal escarpment, the amphitheater pattern, and many medium- and small-scale landforms reflect more complex controls by the interplay of erosion, bedrock resistance, and tectonics (Peulvast et al. 2008).

2.1 The Basement

2.1.1 Structural Framework

The Borborema structural province is composed of "large tracts of Paleoproterozoic gneisses and migmatites surrounding several smaller nuclei of Archean crust" (Santos et al. 2008). These high-grade metamorphic rocks, formed at infrastructural levels and often affected by partial melting, are overlain by remnants of Paleoproterozoic to Neoproterozoic supracrustal rocks, and all of them are intruded by ubiquitous Neoproterozoic granitoids. The province is subdivided into tectonic blocks by branched, anastomosed EW- and NE-trending shear zones which form a mechanical coherent system over more than 200,000 km^2, one of the largest lithospheric transcurrent shear zone systems in the world (Vauchez et al. 1995; Fig. 2.1). This system, "a gigantic, dextral strike-slip duplex" (Alkmim et al. 2001), belongs to the Brasiliano/Pan-African orogen, the last of the Proterozoic collages which began with the closure of the large Pharusian-Goias ocean between the Parnaiba and Borborema blocks at 620–600 Myear and progressively consolidated the West Gondwana (Almeida et al. 2000). This phase of generalized crust thickening marked by the development of eclogites and high-temperature thrusting foliation was followed by the dextral northward extrusion of the Borborema province along the set of shear zones during the final closure of the Sergipano ocean between the West Africa and São Francisco-Congo cratons (590–580 Myear: Araujo et al. 2013).

The EW-trending Patos shear zone, in Paraiba and southern Ceará, divides the basement of the study area into two domains: the Northern Domain, to the north, and the Transverse Zone domain, to the south (Ferreira et al. 1998a). According to these authors and to Vauchez et al. (1995), at least six crustal blocks are recognized in the Northern Domain, five of which are represented here, from NW to SE: the

Médio Coreaú, Central Ceará, Schist Belt, Rio Piranhas, and Seridó tectonostratigraphic units (Fig. 2.1). The last three units correspond to the Rio Grande do Norte block of Santos et al. (2008). In most of these, Proterozoic cover sequences represented by former cratonic to pericratonic sediments, locally of rift type, rest upon a Paleoproterozoic continental basement (Caby et al. 1995; Monié et al. 1997).

Between the Sobral-Pedro II (or Transbrasiliano) and Senador Pompeu shear zones, in the 200-km wide *Central Ceará domain*, the basement is formed by monotonous orthogneisses of tonalitic to granodioritic composition, migmatitic granites, and migmatites, and remnants of amphibolites and metasediments from former greenstone belts (southern Ceará), of Paleoproterozoic age, but reworked from a mostly Archean reactivated basement (Caby and Arthaud 1986; Monié et al. 1997; Fetter et al. 1999). Better preserved Archean terranes are identified within this domain. In its southern part, between the Farias Brito Shear Zone and the Patos Lineament, the Granjeiro Unit (2.55 Ga: Arthaud et al. 2008) is made of strongly deformed plutonic rocks (tonalities, granodiorite) cross-cutting mafic metavolcanic rocks and associated metasediments. In central Ceará, the Troia-Pedra Branca massif (Santos et al. 2008), corresponding to the Mombaça and Cruzeta complexes of Arthaud et al. (2.8 and 2.7 Ga) is mostly made of migmatitic gneiss with compositional layering (granodiorite bands alternating with tonalite and granite bands). The overlying cover rocks of the Ceará Group are pelitic to semipelitic schists and paragneisses of high-grade amphibolite facies and Mesoproterozoic age, more or less mobilized and migmatitized, and interlayered with quartzites, marbles, and calc-silicates, with gneisses derived from former sills and laccoliths, and with thick sheets of leucocratic granitoids. They are interpreted as a passive margin-type sedimentary unit, the sedimentation of which began around 750 Myear, following rifting of the Archean/proterozoic basement (Arthaud et al. 2008, 2014).

All these structures have been affected by later development of Himalayan-type plurikilometric nappes (Independência nappes) with westward to southwestward vergence accompanied by reverse metamorphism (presence of migmatites in the upper part of the nappes) and involving both the Archean basement and the Proterozoic units. These nappes are well-identified west and southwest of Fortaleza (Maranguape and Uruburetama klippes) and in central Ceará, where the supracrustal rocks (amphibolites, meta-arkose, quartzite, calc-silicate rocks) of the Algodões Unit (Arthaud et al. 2008) lie with a flat retrogressive sole thust upon the less deformed and non-migmatitic foreland (the Troia-Pedra Branca inlier of Archean crust: Santos et al. 2008). To the west, the gneisses of granodioritic and granitic origin of the Tamboril-Santa Quiteria Complex may represent a vast batholith (40,000 km^2) intruded in late- to post-tectonic continental magmatic arc of Neoproterozoic age and its fore-arc and back-arc sediments, volcanic and volcaniclastic rocks (665–622 Myear; Fetter et al. 2003; Arthaud et al. 2008). This structure requires the subduction of oceanic crust and implies that an oceanic domain existed in the NW of the Borborema Province prior to the amalgamation of West Gondwana (Santos et al. 2008; Amaral et al. 2011).

Northwest of the Transbrasiliano Lineament or Sobral-Pedro II shear zone, a strike-slip belt reactivating previous sutures (Araujo et al. 2013), the *Médio Coreaú domain* is correlated with the more external part of the Pan-African belt exposed east of the Pan-African suture in Ghana, Togo, and Benin (Monié et al. 1997). To the north, it comprises deformed rocks of Paleoproterozoic origin, mainly represented by the Granja granulite belt (orthogneisses, metapelitic rocks, mafic granulites affected by retrogressive metamorphism during the Brasiliano orogeny). Locally bearing the marks of a Late Paleoproterozoic extensional event (felsic to intermediate volcanic rocks of the Saquinho Volcanic Sequence), this basement complex is unconformably covered by two main supracrustal sequences: the Neoproterozoic Martinópole and Ubajara groups. The oldest (Martinópole Group, 775 Myear) is formed of fine-grained sedimentary rocks (schists, quartzites) deposited in an intracontinental rift evolving to marine conditions and then metamorphosed to greenschist or amphibolites facies during the Brasiliano orogenesis (Santos et al. 2008). South of Coreau and of the Arapa fault, the Ubajara Group, mainly made of metamorphosed slate, siltstone, sandstone, and carbonates, represents a proximal, stable-shelf sedimentary sequence. To the south, this metamorphic sequence is adjacent to a Late-Brasiliano (Early Paleozoic) molassic or rift basin (the narrow Jaibaras trough, between the Café-Ipueiras fault and the Transbrasiliano Lineament: Oliveira and Mohriak 2003). These last units are described further (Chap. 3), together with the Brasiliano and Late-Brasiliano structures to which they are related, as well as the latest of the granitic intrusions that intrude the basement rocks and the sedimentary sequences. One of these granites (the porphyritic granite of Chaval, in the NW of the domain) is slightly older (591 ± 10 Myear: Fetter et al. 2000), and was affected by the Brasiliano deformation.

To the southeast of the Senador Pompeu shear zone, along which the Quixadá-Quixeramobim synkinematic batholith and the Mombaça granulites are exposed, the *Schist Belt domain* (part of the Rio Grande do Norte block) displays a linear trend on hundreds of kilometers, with spectacular bends. It is made of an association of antiformal gneiss cores representing the pre-Brasiliano basement rocks, and of synformal schist belts of the Proterozoic cover, that have only undergone low-pressure metamorphic conditions (Monié et al. 1997). Between the Óros and Jaguaribe shear zones, the *Orós-Jaguaribe* belt is made of two subparallel, sinuous, and narrow (40 km in the south, 4 km in the north) synformal quartzitic, calcareous and metapelitic zones derived from the former sedimentary and volcanic rocks formed in an intracontinental rift 1.79 Ga ago (Fetter et al. 1999). This belt is preserved in a Brasiliano structure which does not correspond to a deep lithospheric discontinuity (De Castro and Costa 1999), but has played an important role in the subsequent evolution. It separates through a steep tectonic contact, the westerly migmatitic domain from a wide easterly *Rio Piranhas domain*, composed of polycyclic pre-Brasiliano basement gneisses (Vauchez et al. 1995), and severely intruded by Brasiliano granitoids. To the east, in the *Seridó belt*, the lower pre-Brasiliano granitic-gneissic basement (Caicó complex) crops out as elongated NNE–SSW domes and synformal metasedimentary units (carbonates, metapelites, and quartzites) between belts of the younger Neoproterozoic cover (Jucurutu, Equador, and Seridó Formations, <650 Myear; Caby et al. 1995; Neves 2002; Arthaud et al. 2008).

South of the 20–30-km wide mylonitic and migmatitic Patos dextral shear zone, a probable suture of Neoproterozoic age running EW for 300 km (Costa et al. 2002), the *Transverse Zone domain* is represented in the study area by several terranes in which late Mesoproterozoic supracrustal belts alternate with blocks of Archean to Paleoproterozoic high-grade gneisses and migmatites. They are widely intruded by numerous Brasiliano plutons of varied compositions (Ferreira et al. 1998b).

2.1.2 Brasiliano and Late-Brasiliano Structural Heritage

The structural pattern of the basement is mainly inherited from the Brasiliano orogeny. In the study area, this orogenic cycle is the result of an end-Proterozoic, probably oblique, convergence between the West African-São Luis Craton, in the west, and the Congo-São Francisco Craton, in the southeast, following a rifting stage recognized in western Ceará around 800 Myear (Martinópole and Independência Groups: Fetter et al. 1999). This phase of crustal stretching would have been controlled by slab pull forces generated by outwardly dipping subduction zones, with formation of oceanic crust after ca 700 Myear, only to be subducted shortly afterward (Neves 2015). Therefore, the Borborema Province is mainly the result of a dominantly intracratonic evolution. Trench retreat and slab rolling back of the plate would have led to collision of the different terranes of the Transbrasiliano Belt with the passive margin of the paleocontinent, giving rise to the beginning of contractional deformation. 880–800 Myear old rocks with juvenile characteristics recently discovery in the Tamboril-Santa Quiteria Complex of the Ceara Domain (Araújo et al. 2014) could be slices of arc rocks transported southwards.

The main tectono-metamorphic events (620–580 Myear) generally culminated in the amphibolite facies (Monié et al. 1997). They were followed by the intrusion of a large volume of syn- to late-kinematic granitoids, by late-orogenic transcurrent movements under decreasing temperature conditions until 500 Myear, and by opening of molassic grabens in Ceará (Trompette 1994; Almeida et al. 2000). Earlier cooling occurred in the outer part of the belt, especially in the northern Medio Coreau domain by 575–560 Myear, whereas to the south of the Senador Pompeu shear zone, the inner part experienced cooling later, in the range 550–500 Myear (Monié et al. 1997).

Many important morphological features are controlled by the huge ductile lithospheric faults which cross-cut the region. These faults belong to a vast system where, in the pre-drift situation, "EW-trending lineaments, more than 1000 km long from NE Brazil to Cameroon, merge into NE-trending lineaments that run about 3000 km from central Brazil to Hoggar" (Vauchez et al. 1995). Steeply dipping, they are typically 1–10-km wide (up to 30 km for the Patos shear zone) and comprise mylonites, migmatites, granulites (e.g., the Mombaça granulite belt) and syn- to late-kinematic granitic plutons and dykes. To the west (Ceará), this system comprises NE–SW shear zones which postdate nappe emplacement (the Granja,

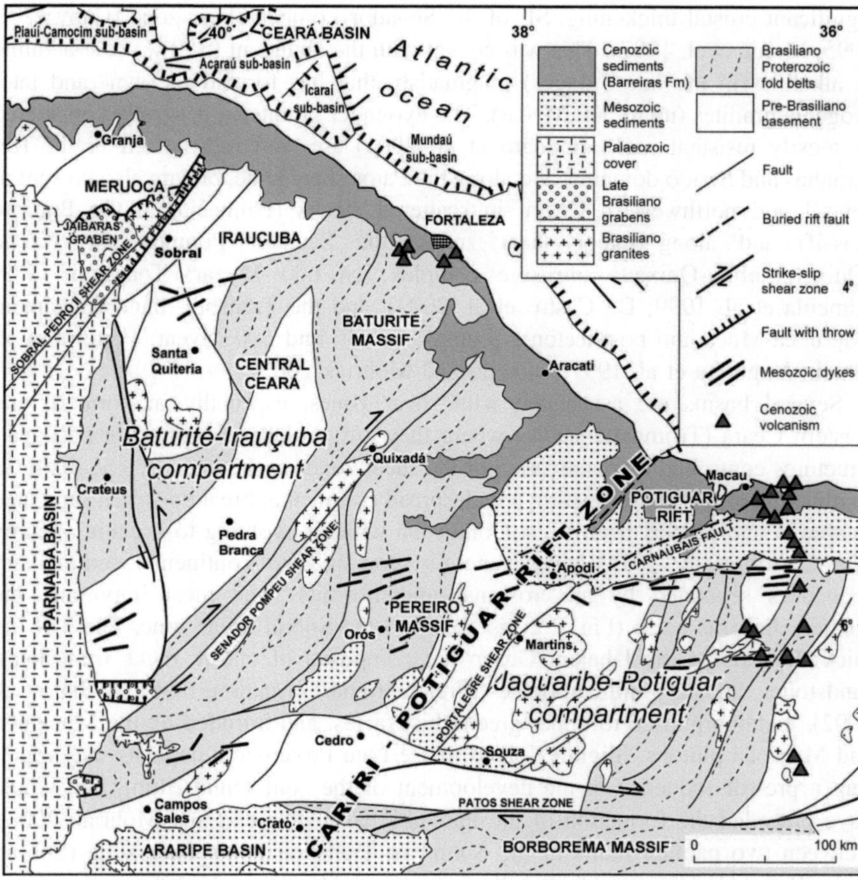

Fig. 2.2 Geological map of the study area, compiled after Brito Neves et al. (2000), Carneiro et al. (1989), and Caby et al. (1995)

Sobral-Pedro II or Transbrasiliano, Senador Pompeu-Quixeramobim, Tatajuba, and Potengi shear zones; Fig. 2.2), and along some of which megafolds and sigmoidal structures locally affect the Mesoproterozoic cover (Tatajuba shear zone near Pio IX: Oliveira and Maas 2002). In the east (mainly Paraiba and Rio Grande do Norte states), sinuous NE–SW shear zones (Jaguaribe, Portalegre) and transpressional belts—the Seridó and Cachoeirinha belts—branch on the EW Patos system. Tectonic activity along these shear zones took place at shallow depths until the Cambrian-Ordovician, and perhaps until the Devonian or later along the Transbrasiliano lineament (tilted and sheared Devonian sandstones near Santana do Acarau; Arthaud et al. 2008).

The mylonitisation began around 600–570 Myear, in continuity with the previous deformation, forming crystalline nappes and a thickened crust in northern and central Ceará, and upright folding in a dextral transpressional regime without significant crustal thickening, SE of the Senador Pompeu shear zone (Caby et al. 1995; Monié et al. 1997). This was coeval with the abundant (>50 %) calc-alkaline to alkaline (in the latest stages) magmatism that has formed the syn- and late-orogenic granites (up to 500 Myear). The «younger granites», generally considered as mostly resistant rocks (Godard et al. 2001) are well represented in the Rio Piranhas and Seridó domains, and along the Patos shear zone, but are also present in central and northwestern Ceará, in scattered stocks (Palmacia, in the Baturité massif) and along major shear zones: the Senador Pompeu shear zone (Quixeramobim-Quixadá intrusive complex, ca 600 Myear: Torquato 1989; Almeida et al. 1999; De Castro et al. 2002), and the Transbrasiliano Lineament (Meruoca-Mucambo post-tectonic plutons, ca 540 and 500 Myear: Ferreira et al. 1998b; Nogueira et al. 1997; Oliveira and Mohriak 2003).

Several basins are associated with shear zones, especially in northern and western Ceará (Trompette 1994), where they are partially preserved in pull-apart structures controlled by reactivation of the shear zones in the interval 0.58–0.50 Ga (Arthaud et al. 2008). Reflecting a widespread continental breakup which separated Laurentia and Baltica from West Gondwana without evolving to oceanic opening (De Castro et al. 2014), they generally contain two continental sedimentary sequences separated by an erosional unconformity. The most important, the Ubajara-Jaibaras basin (Fig. 2.2), is linked to the Sobral shear zone. The 3–4 km thick deposits of the Ubajara Group are composed of clastic rocks (red slates, sandstones, arkoses) and carbonates (Freicheirinha Formation, 0.98 Ga: Sial et al. 2002), metamorphosed to upper greenschist facies, and intruded by the Mucambo and Meruoca plutons. Slightly folded, these Late Proterozoic deposits may represent a pre-rift sequence in the development of the continental rifting that would have preceded the formation of the Parnaiba basin (Oliveira and Mohriak 2003). Between two parallel faults of the Sobral or Transbrasiliano shear zone (Sobral-Pedro II and Café-Ipueiras faults), they are covered by the anchimetamorphic, 3-km thick, clastic fluvial deposits of the Jaibaras Group, which are also intruded by the Meruoca granite, probably emplaced in a pull-apart structure (Teixeira et al. 2004), and by other magmatic rocks (sills and dikes of the Parapuí suite). Considered by Destro et al. (1994) as late-orogenic molassic deposits of fluvial origin formed in a narrow graben (5–10 km), they include conglomerates and sandstones (the Massapê basal unit, Vendian-Cambrian), fine sandstones, siltstone, and shales (the Pacujá Formation, of fluvial to lacustrine origin), and finally conglomerates with clasts of plutonic/volcanic rocks (Mucambo/Meruoca plutons, Parapuí suite). This structure is interpreted by Oliveira and Mohriak (2003) as an aborted rift branch developed in a previously thickened lithospheric-crustal system after the Brasiliano orogeny, between the West African and Brazilian cratonic nuclei, and slightly inverted, before the sag subsidence of the intracratonic Parnaiba basin in the Silurian.

Other occurrences of coarse continental clastic and volcanic-sedimentary rocks are known in grabens of northern Ceará (Sairi-Jaguarapi), eastern Ceará (Sitiá) and southern Ceará (Cococi, Catolé-São Julião, Barro; Teixeira et al. 2004; Fig. 2.2).

Large parts of the orogen were already deeply eroded and even planated at the onset of the Silurian-Devonian post-orogenic sedimentation. The influence of their structures on the present morphological pattern is only indirect. Later faulting was strongly controlled by zones of weakness (Maia and Bezerra 2014a), whereas the complex tectonic and lithological patterns of this orogen allowed a generalized contribution of differential erosion to landform shaping.

2.2 Post-orogenic Rocks and Structures: Pre-rift Deposition and Magmatism

2.2.1 The Post-orogenic Sedimentary Cover

The basement sedimentary cover is geographically and stratigraphically discontinuous. The distribution of its elements reflects strong tectonic and geomorphic controls. The main sequences are Paleozoic deposits of a large sedimentary basin—the Parnaiba Basin—, the syn- and post-rift deposits of the Mesozoic Cariri-Potiguar rift zone and of the transform margin, and a thin coastal fringe of Cenozoic sediments.

The study area comprises the eastern edge of the Parnaiba Basin (Serra da Ibiapaba), a large sedimentary basin centered on a set of Cambrian-Ordovician grabens formed during the last stages of the Brasiliano orogeny (Almeida et al. 2000; Oliveira and Mohriak 2003; De Castro et al. 2014). Its basal layers—fluvial to coastal or neritic conglomerates and sandstones of the Ordovician-Devonian Serra Grande Formation, a few hundreds of meters thick—, grade upwards to finer sandstones. They are flat-lying or slightly tilted westwards, and lie unconformably on the eroded faulted structures and intrusions related to the Sobral shear zone. The metamorphic or igneous substrate is often deeply weathered, as shown by thick kaolinized horizons outcropping below the Ibiapaba escarpment (e.g., Ubajara, Ipu; Fig. 2.3a). The abrupt erosive edge of the basin and the presence of remote ouliers, mainly in NW Ceará (graben of Santana do Acarau, on the Sobral shear zone: Destro et al. 1994; Galvão and Da Silva 2002) and southern Ceará, suggest that the initial size of the Parnaiba Basin was much larger and was reduced by erosion at later times. In southern Ceará, the Cariri Formation is considered by some authors as a counterpart of the Serra Grande Formation outside the Parnaiba basin (Hegarty et al. 2002; Assine 2007). Its quartz conglomerate and coarse sandstone layers (former alluvial fans and braided streams), as well as the overlying deposits of the Late Jurassic (Callovian-Tithonian) pre-rift sequence of the Araripe basin (Brejo Santo and Missão Velha Formations), fill an irregular basement topography. They are preserved in small faulted basins located in the Patos shear zone (Lavras da Mangabeira, Barro) and also outcrop at the periphery of the Araripe Basin (Matos 1992), below the Cretaceous deposits (Nova Olinda, Juazeiro do Norte, Milagres…). On the contrary, such sediments are not found below the Cretaceous deposits of the other basins of the Cariri-Potiguar rift

Fig. 2.3 **a** Ipu (between Sobral and Crateus, south Ceará): the Serra da Ibiapaba, a sandstone glint (Serra do Grande Formation) overlooking the strongly weathered and excavated metamorphic basement, as seen to the south. Slightly dissected pediments below the high sandstone ledge. **b** The north spur of the Serra de Portalegre. The *top* of this dissected mesa corresponds to a thin layer of lateritized fluvial sandstone (Serra do Martins Formation). *Photographs* J.P. Peulvast

zone and of the margin (Matos 1992, 2000), and more elevated regions might have existed to the north of the Patos shear zone.

Many deposits considered as pre-Cretaceous in age rest on a deeply weathered and kaolinized basement with a typical etch morphology. The same was observed for similar flat-lying sediments (coarse quartz conglomerate, sandstone and siltstone layers with oblique stratifications of fluvial origin, coming from distant regions of the basement in the south: Menezes and Lima Filho 1997; Menezes et al. 2002) which cap the small Portalegre, Martins and Santana mesas, south of the Potiguar rift (e.g., north of Portalegre: Fig. 2.3b).

Classically dated as Cenozoic on the basis of lithostratigraphical analogies with mid-Tertiary sediments of the margin (Tibau and Guamaré Formations of the Potiguar Basin; Mabesoone et al. 1972; Saadi and Torquato 1992; Jardim de Sá et al. 1999), and considered by many authors as a key deposit to understanding the morphotectonic history of the Borborema massif (Ab'Sáber 2000; Morais Neto and Alkmim 2001), this so-called Serra do Martins Formation presently occurs as narrow erosive remnants in a culminating position, without any identified source area. Its deposits, a few meters to 50-m thick, are locally intruded by a Late Oligocene (25 ± 1 Myear) basaltic plug (Serrote Preto, Serra De Santana: Morais Neto et al. 2002) and are, therefore, older than Late Oligocene. Deeply weathered and capped by thick laterites, lacking fossils, they could be considered as older than is generally accepted, corresponding to Eocene (Jardim de Sá et al. 1999), Paleocene (up to 64 Myear: Morais Neto et al. 2008) or even older deposits (Peulvast et al. 2008). A diachronous post-Ordovician, pre-Cretaceous sedimentary cover exists or has probably existed in several parts of the area, belonging to the Parnaiba Basin or to smaller basins whose nearby source areas might have been paleo-fault scarps and local uplifts related to late-orogenic or pre-rift movements (Martill 1993). It had already been eroded before the Cretaceous, or had never been deposited, in the areas presently occupied by the continental margin (Matos 2000) and by most continental rifts north of the Patos shear zone.

2.2.2 *Early Magmatism and Rifting*

Whereas the final rifting and continental separation occurred during the Aptian, Albian, and Cenomanian (Matos 2000), deformation associated with the South Atlantic opening began earlier, during Jurassic-Triassic time (225–160 Myear), accompanied by early extension and magmatic activity within the Gondwana supercontinent (Maurin and Guiraud 1993; Guiraud and Bellion 1995). The widespread fluvial and shallow sedimentation of Late Jurassic age known in the Recôncavo-Tucano-Jatobá, Sergipe-Alagoas, and Araripe basins is interpreted by some authors as recording the early phase of crustal stretching (Matos 1992; Martill 1993).

Pre-rift magmatic activity is only known north of the Patos shear zone, with a roughly EW array of mafic dykes (olivine-free diabases)—the Rio Ceará Mirim Dyke Swarm—which may be related to this early phase of crustal stretching. This swarm was intruded in the basement of the Borborema Province during two main pulses around 145 and 130 Myear (Tithonian-Hauterivian; Sial et al. 1981; Almeida et al. 1988; Oliveira and Gomes 1996; Mizusaki et al. 2002) or 134–110 Myear (Hauterivian-Aptian: Hollanda et al. 2006). This 400-km long tholeiitic dyke swarm is mainly identified from the southern border of the Potiguar basin to the Jaguaribe valley between the Iguatu and Icó basins, widening from 30 to 100 km. Probably associated with the initial stages of the formation of the Potiguar basin (Françolin and Sztamari 1987), it forms an arcuate array trending from N90°E (northeastern Borborema Province) to N55°–70°E in the southwest, where the dykes become nearly parallel to the NE-trending Eocretaceous rift system (Matos 1992; Fig. 2.4).

2.3 The Continental Rift

The occurrence of well-dated Mesozoic and younger sedimentary basins in the northern Nordeste is restricted to the rift structures and the transform margin formed during the Cretaceous as a result of the separation between Africa and South America. Therefore, they are found in the intracratonic Cariri-Potiguar rift zone or depression (Matos 1992; Valença et al. 2003), an aulacogen of Early Cretaceous age,

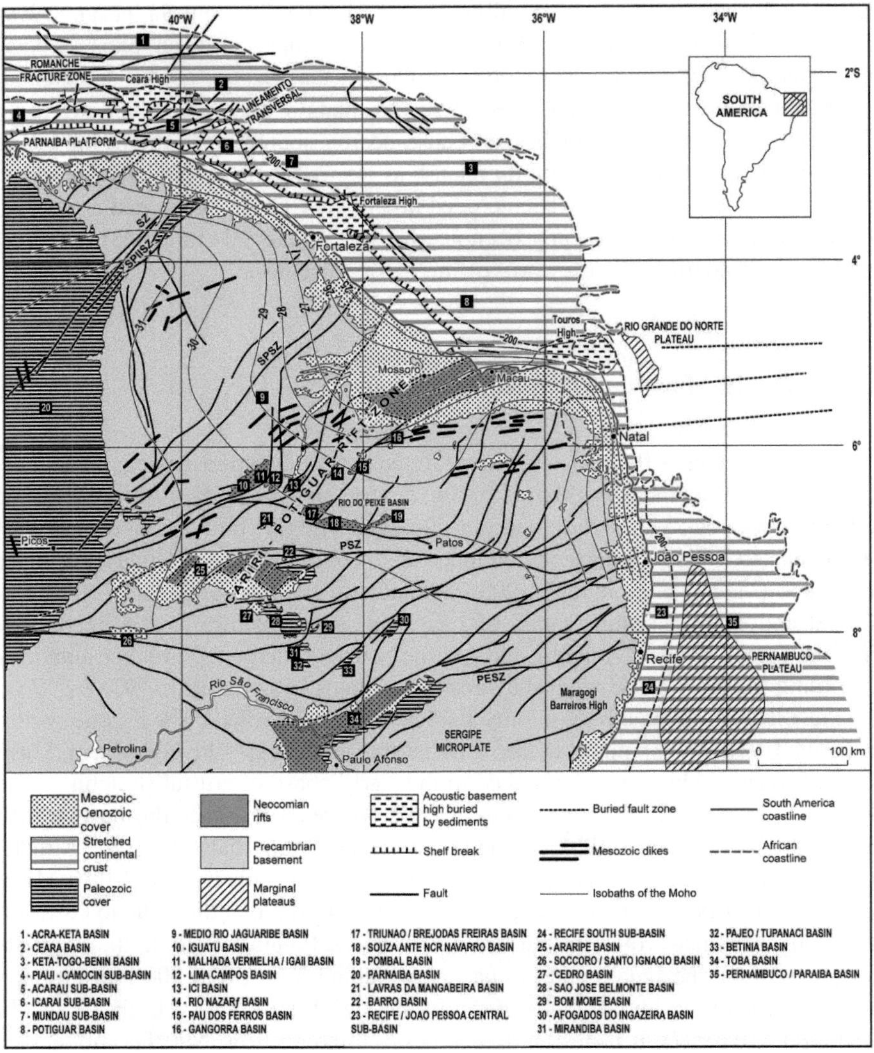

Fig. 2.4 Structural map of the Borborema province, showing the pre-, syn-, and post-rift structures along the Cariri-Potiguar rift zone. Compiled from Matos (1992, 2000) and Valença et al. (2003)

and along the Equatorial Atlantic margin, whose formation was initiated by almost instantaneous oblique extension during the Aptian stage (Matos 2000). According to this author, the intracontinental and offshore basins of the study area can be grouped into two ages: Neocomian-Barremian and Aptian. The location and architecture of the first group are strongly controlled by upper crustal heterogeneities—the shear zone systems of the Borborema province—, whereas the second group was not significantly influenced by this heritage, indicating different processes of lithospheric thinning.

2.3.1 Limits of the Rift Zone

According to Matos (1992), the Cariri-Potiguar basins typically consist of asymmetric half grabens separated by basement highs, transfer faults, and/or accomodation zones. Most of these are a few tens of kilometers wide with a NE–SW trend, perpendicular to the extension direction. Master normal faults are often controlled by Proterozoic shear zones (Orós, Jaguaribe, Portalegre: De Castro et al. 2012), but the Rio do Peixe Basin was also formed in left-lateral transtensional conditions along the EW Patos shear zone (Françolin et al. 1994). Together with minor isolated basins (Tarrafas, Pau dos Ferros, Nazaré, Gangorra), they are grouped in a 150 km wide, 500-km long NE–SW zone, from the offshore Potiguar basin to the Araripe basin (Fig. 2.4). Most of their asymmetric structures dip to the SE, like the wide basement blocks which form their shoulders, the Aracati Platform-Central Ceará highlands to the NW, and the Touros High-Borborema to the SE.

Since the presently observed basins mainly represent erosional remnants, the timing of the last stages of sedimentation and rifting processes as well as the initial limits of the rift zone are not precisely known, but the shallow topography of the basement on both sides precludes the former existence of other deep basins outside the present limits. However, the structural control of the southern edge of the Parnaiba basin by the SW continuation of the Senador Pompeu shear zone (Carneiro et al. 1988) suggests some post-Paleozoic reactivation of this fault zone, which could have belonged to *en échelons* systems of antithetic border faults reinforcing the rotation of the hanging wall block on the NW margin of the Cariri-Potiguar rift zone (Fig. 2.5).

Faulting and tilting of basement blocks related to passive preopening stretching may have occurred in a wider region, outside the Cariri-Potiguar rift, as suggested by the width (300–400 km and more) of the rifting zone south of the study area. In spite of the lack of Mesozoic sediments, faulting related to Cretaceous rifting may be inferred along a few Proterozoic shear zones. The most probable case of reactivation is recorded on the Sobral-Pedro II shear zone (NW Ceará) by the intense, transpressional deformation of an early Devonian sandstone body preserved at low altitude as an outlier of the Parnaiba basin in the Santana do Acaraú graben, about 30-km long and 4-km wide (Destro et al. 1994; Galvão and Da Silva 2002). The age of the reactivation is poorly constrained. According to Arthaud et al. (2008), it might correspond to the continuation of tectonic activity along the Transbrasiliano lineament during the Devonian. However, the identification of similar structures along the same fault zone in the Parnaíba Basin (Serra da Ibiapaba north of Ipu: Fig. 2.4), and in Aptian to early Cenomanian sediments of the offshore Ceará basin (Costa et al. 1990; Matos 2000) suggests that dextral transpressional reactivation might have occurred in the mid-Cretaceous, possibly related to transform faulting and opening of the Equatorial Atlantic. Downfaulting of the sandstone body along normal faults of a probable pull-apart basin (Galvão and Da Silva 2002), which allowed its preservation from erosion, has occurred earlier, in the Late-Devonian, or during the Early Cretaceous intracontinental stretching phase.

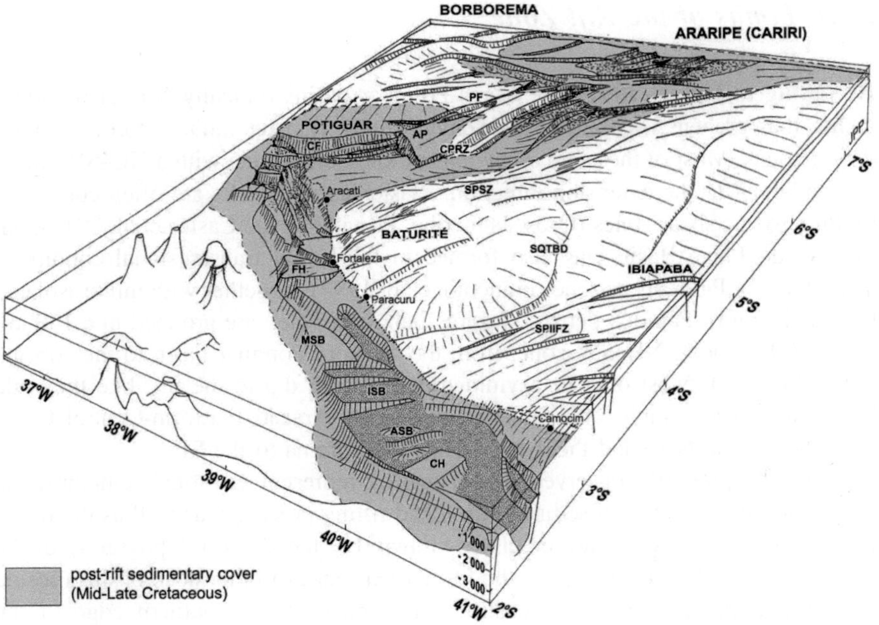

Fig. 2.5 Sketch block diagram of the study area showing the Mesozoic tectonic pattern such as it would hypothetically look like without erosion and post-rift deposition. *Dotted areas* syn-rift deposits. *AP* Apodi transfer fault; *ASB* Acaraú sub-basin; *CF* Carnaubais fault; *CH* Ceará High; *CPRZ* Cariri-Potiguar rift zone; *FH* Fortaleza High; *ISB* Icaraí sub-basin; *MSB* Mundaú sub-basin; *PF* Portalegre fault; SPIIFZ: Sobral-Pedro II fault zone; *SPSZ* Senador Pompeu shear zone; *SQTBD* Santa Quiteria-Tamboril-Baturité Domain

The limits of the onshore area of crustal thinning related to NW–SE extension in the Cariri-Potiguar rift zone (from 32- to 26-km in thickness: De Castro and Medeiros 1997; De Castro et al. 1997, 2012) do not include the narrow Araripe basins, to the southwest. Therefore, maximum crustal stretching occurred north of the Patos shear zone, in a wide and short area around the Potiguar basin, between the Touros High and the Borborema in the southeast, and the Senador Pompeu shear zone, which forms the limit of the crustal thickening zone of the Central Ceará highlands in the northwest (Fig. 2.4).

2.3.2 The Cariri-Potiguar Rift Zone

In southern Ceará, the Araripe Basin is presently uplifted to 700–900 m above sea level and, due to Late- to Post-Cretaceous regional uplift and topographic inversion, overlooks the surrounding basement areas (Peulvast and Bétard 2015). Below the post-rift cover, the Araripe rift basin is divided into two sub-basins (W: Feitoria or Feira Nova sub-basin; E: Eastern or Cariri sub-basin; Ponte and Ponte-Filho 1996;

De Castro and Castelo Branco 1999; Martill et al. 2007; Assine 2007) separated by the Dom Leme crystalline horst. Inside, numerous NE–SW grabens filled by rift sediments are separated by minor horsts, transfer faults, and/or accommodation zones formed along preexistent sigmoid shear zones connected to the north and the south with the major E–W-trending Patos and Pernambuco shear zones (Corsini et al. 1991; Matos 1992). They contain up to 1800 m of Late Jurassic and Early Cretaceous fluvio-lacustrine sediments lying on the pre-rift series.

Before and during the rifting, fluvial and lacustrine deposits of Jurassic and Early Cretaceous age (Vale do Cariri Group) were deposited upon the sandstone cover of the basement (Mauriti or Cariri Formation, Paleozoic; Martill et al. 2007; Assine 2007; Fig. 2.6). Following a Late Jurassic-Early Cretaceous phase of regional subsidence and continental sedimentation (Brejo Santo and Missão Velha Formations: clay, siltites, sandstones, up to 600-m thick; Coimbra et al. 2002; Valença et al. 2003), the main rifting stage (Berriasian-Early Barremian) was that of graben infilling by thick fluvial, deltaic, and lacustrine sediments transported to the SSW (Abaiara Formation: siltites, sandstones, and conglomerates). In the same time, the intervening basement areas and the overlying sediments were eroded, and then partially beveled during the transitional post-rift stage (Early Aptian), except for some residuals ultimately buried below post-rift series and then exhumed (Horto, Serra do Juá). Along the faults limiting the Crato-Juazeiro do Norte graben, throws and sediment thicknesses reach up to 1000 m (Assine 2007).

Beyond the small isolated grabens of Barro and Lavra da Mangabeira, filled with pre-rift sandstones, and the narrow Tarrafas basin, aligned on the Tatajuba shear zone and filled with Neocomian sediments, two groups of basins are found 80 km to the N and NE. Along the Jaguaribe valley, the Iguatu-Malhada Vermelha-Icó-Lima Campos group consists of NE-trending shallow half grabens, 5–20-km wide, with listric normal faults on their SE sides. Controlled, as the Potiguar basin, by reactivated basement shear zones (De Castro et al. 2008), they are filled with Neocomian continental conglomerates, sandstones, siltstones, shales, and rythmites (Matos 1992). To the east, the Pereiro horst zone, including a narrow NE-trending graben (Icózinho), separates them from the Rio do Peixe basin. From W to E, this basin is made of four adjacent sub-basins (Triunfo, Brejo das Freiras, Sousa, Pombal: Sénant and Popoff 1991; Françolin et al. 1994), separated from each other by small basement highs and aligned on the EW Patos shear zone which forms the southern limit of the system (De Castro et al. 2007; Nogueira et al. 2015). They correspond to SW–NE or EW (Sousa) trending half grabens, 5–15-km wide, and contain up to 1000–1500 m of syntectonic fluvial and lacustrine sediments of Berriasian to Aptian or Early Albian age. From the base to the top, these deposits comprise conglomerates, sandstones, siltstones, shales, and white calcareous lenses. They are downfaulted and tilted to the SE along the Malta, Portalegre, and Rio Piranhas faults. To the north, the small Nazaré, Pau dos Ferros, and Gangorra basins contain similar deposits, finding their source from nearby fault escarpments (Srivastava et al. 1989; Corsino and Tiriba 1985; Fig. 2.4).

The largest intracontinental basin in the study area is the Potiguar basin, which extends both onshore and offshore in Ceará and Rio Grande do Norte, from about

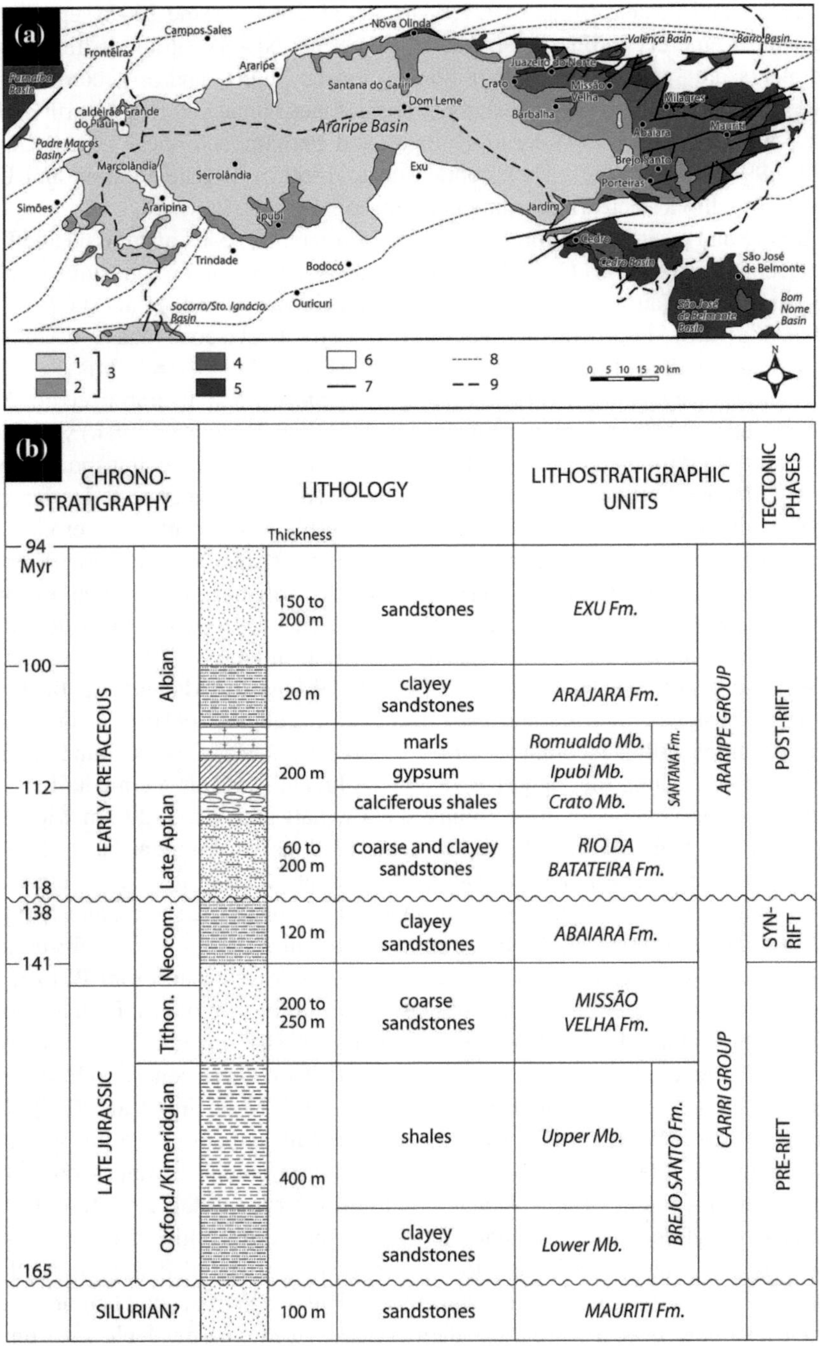

Fig. 2.6 Geological map and stratigraphic log of the Araripe-Cariri region. **a** Map modified from Ponte and Ponte-Filho (1996), Assine (1994). *1* Exu Formation; *2* Arajara, Santana, Rio da Batateira Formations; *3* Araripe Group; *4* Cariri Group; *5* Mauriti Formation; *6* basement complex; *7* fault; *8* lineament; *9* state border. **b** Log compiled from Ponte and Ponte-Filho (1996), Neumann (oral comm. 2008)

200-km inland to the present continental shelf and slope, at least to water depths of 2000 m (Bertani et al. 1990). Reaching 6000 m in thickness, it covers 48,000 km^2 (including 21,000 km^2 onshore). The Early Cretaceous rift is entirely covered by the post-rift series of Late Cretaceous age which form the Chapada do Apodi (Fig. 2.7). Below a thin transitional sequence of Aptian-Eo-Albian age, it contains up to 4 km of syn-rift fluvial, deltaic, and lacustrine sediments of Early Neocomian to Early Barremian age. These deposits—mainly sandstones and siltites—are found in two main NE–SW trending half grabens (Umbuzeiro, Boa Vista), separated by horsts or half-horsts which were locally beveled by erosion before the deposition of the transitional series (Quixaba, Serra do Carmo highs) (Neves 1989). To the NE, the Macau basement high separates these grabens from the offshore Macau low. The basins are limited to the SE by major listric faults, the Carnaubais (in the NE continuation of the Portalegre fault: De Castro et al. 2012), and Quixaba-Serra do Carmo fault zones, and to the SW by the NW–SE Apodi transfer fault, which corresponds to the southern limit of the extended crust (Fig. 2.7). The main depocenters are concentrated along the Carnaubais fault.

2.3.3 From Continental Rifting to Transtensional Opening of the Equatorial Atlantic

To the NW of the Potiguar basin, the triangular Aracati platform (Fig. 2.5) is a southward tilted portion of the basement, limited on its northeast side by the offshore continuation of the Potiguar basin. Its northern part includes small *en échelon* offshore transtensional grabens, NW–SE trending, bounded by EW transfer faults, and filled with Late Barremian sediments (Jacaúna and Messejana grabens; Fig. 2.4). Whereas, onshore rifting failed in the whole Cariri-Potiguar rift system by the end of Barremian (Matos 2000), the formation of these basins, contemporaneous with the early rifting stages in the Benue trough in Africa, suggests that the major deformation became from that time onwards located on the Equatorial branch of the rift system. The onset of EW extension generated transtensional conditions in the Potiguar basin, and rift-controlled sedimentation in its offshore parts up to the Fortaleza high, accompanied by southward tilting of the Aracati platform and of the onshore rift system (Matos 1992).

Along the future Equatorial margin, this new early Aptian tectonic phase first induced the formation of NW–SE trending *en échelon* sag-type basins or synclines truncated by a diffuse fault pattern, without typical rift structures and associated sedimentary facies (offshore Potiguar Basin, Ceará Basin: Matos 2000; Fig. 2.8). The transition between continental and marine conditions occurred during the lower Albian in most of these basins (Potiguar, Piauí-Camocim, Barreirinhas). A deep oceanic circulation was already organized between the Equatorial and eastern margins of Brazil during the middle Albian (Benkhelil 1988; Koutsoukos 1992), but the definitive open ocean connection seems to have been established during the Cenomanian-Turonian transition, when the last link of continental crust between Brazil and Africa had subsided (Valença et al. 2003).

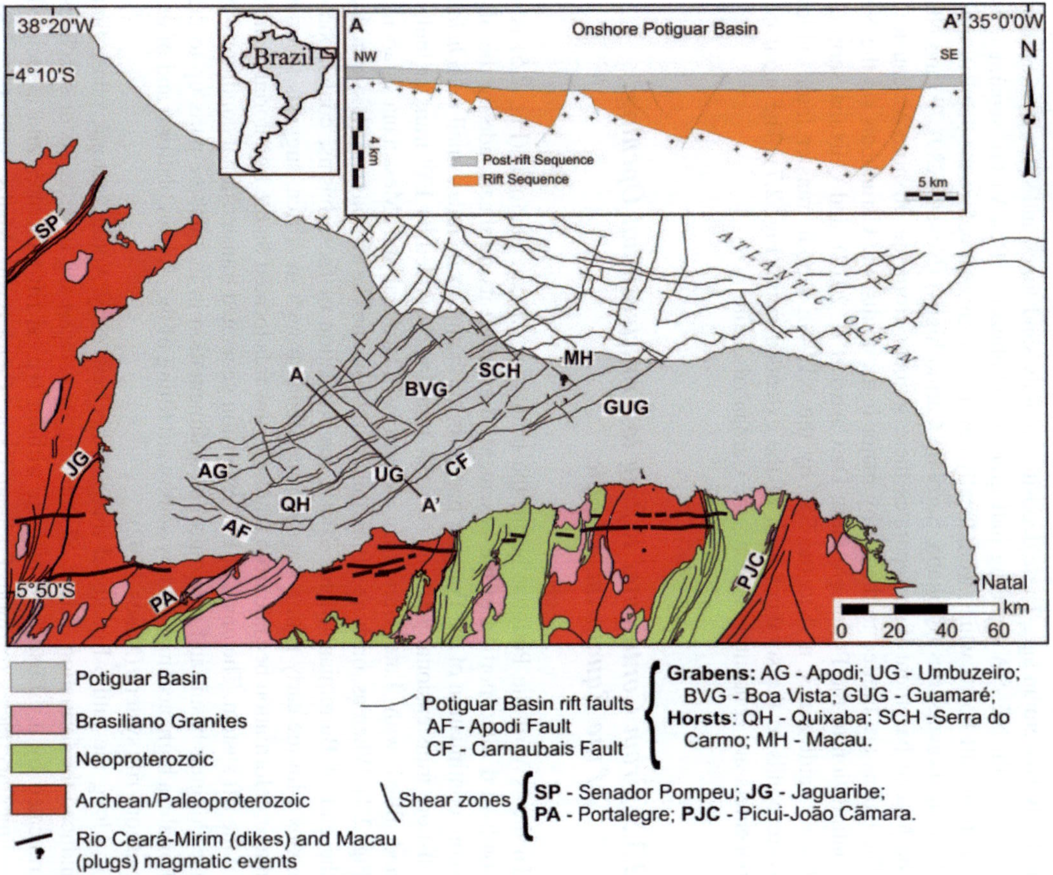

Fig. 2.7 Simplified geologic map of the Potiguar Basin area. From De Castro et al. (2012)

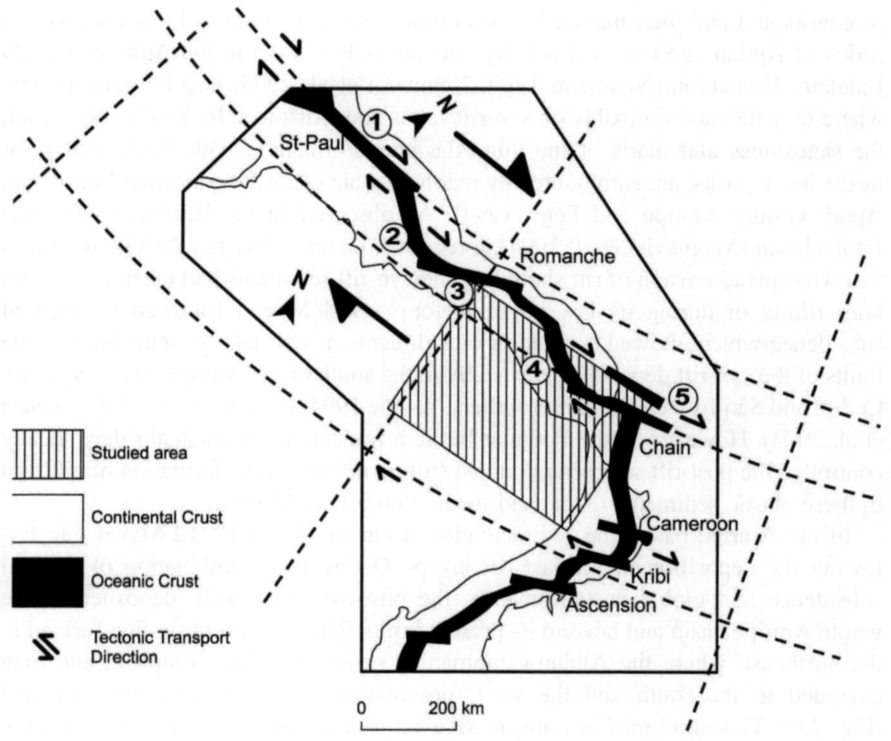

PURE SHEAR-DOMINATED TRANSTENSION	Barreirinhas ; Tanos		2
	Mundau-W Potiguar ; Keta-Togo-Benin		4
WRENCH-DOMINATED TRANSPRESSION	Icarai-Acaraú (Ceará)		
	Sait Pona ; Accra	*Romanche*	3
WRENCH-DOMINATED TRANSTENSION	Para-Maranhão ; Côte-d'Ivoire	*St-Paul*	1
	Potiguar-Benin-Benue	*Cnain*	5

Fig. 2.8 Opening of the Equatorial Atlantic (Albian) and major releasing and restraining bends in the Equatorial shear corridor. Redrawn from Matos (2000)

2.4 The Post-rift Evolution and the Transform Margin

2.4.1 Onshore Post-rift Covers

By the time when transtensional conditions had become prevailing in the Equatorial Atlantic domain, the aborted intracontinental rifts of the Cariri-Potiguar trend had already begun to evolve in post-rift conditions, through thermally induced subsidence (Mello 1989). Until the Cenomanian or Campanian stages, and after a 20 Myear interval of erosion/nondeposition, these rifts and the topographically depressed adjoining areas became locations of widespread sedimentation: first lacustrine or transitional continental/marine, finally continental in the Araripe basin, or continental and then marine in the Potiguar basin. Transitional syn-rift/post-rift series of Aptian age and post-rift deposits are only known in the Araripe (Rio da Batateira Formation: Neumann 1999; Neumann et al. 2003) and Potiguar basins, where they lie unconformably on syn-rift series and structures. In the Potiguar basin, the sandstones and marls of the thin Alagamar Formation, with fluvial-deltaic or lacustrine deposits, are surmounted by marine-deltaic deposits (Alagamar Formation, Apodi Group: Araripe and Feijó 1994). As observed in the Recôncavo-Tucano-Jatoba basin (Magnavita et al. 1994), a regional unconformity (ca 20 Myear) shows that widespread erosion of rift shoulders and syn-rift sediments had taken place soon after rifting or during its last stages, before ~114 Myear, followed by thermal subsidence which allowed deposition of sediments in vast lake systems beyond the limits of the syn-rift deposition, especially to the south of the Araripe basin (Socorro, Cedro, and São José de Belmonte outliers; Assine 1994; Hegarty et al. 2002; Valença et al. 2003). However, in the Potiguar basin, a few basement residual ridges locally controlled the post-rift sedimentation and still participate in the formation of oil traps in these clastic sediments (Açu sandstones: Sousa et al. 2008a).

In the Araripe basin, the sedimentation resumed after a 10–12 Myear gap following the deposition of the last rift layers. Owing to a combination of thermal subsidence and global eustatic events, the post-rift series were deposited in the whole Araripe basin and beyond its present limits. The post-rift basin first formed in the northeast, where the Albian-Cenomanian series are more complete, and then extended to the south and the west, progressively lapping onto the basement (Fig. 2.6). This stratigraphic transgressive–regressive sequence, known as Araripe Group (Ponte and Ponte-Filho 1996) is made of fluvial, lacustrine, lagoonal, and marine sediments of Aptian to Cenomanian age, divided into several formations: Rio da Bateira (or Barbalha: Assine 2007), Santana (Crato, Ipubi, Romualdo Members), Arajara, and Exu.

In the east (Cariri depression), the post-rift series comprises 250–280 m of mainly soft rocks overlain by thick massive sandstones (Exu Formation, 150–250 m). In this area, they offer the best potential to differential erosion. The coarse to fine and clayey sandstones of the basal post-rift layers (Rio da Batateira or Barbalha Formation) correspond to two fluvial cycles with upward decrease of grain size and lacustrine pelitico-carbonated intervals ("Batateira layers": Assine 2007). Reaching 60–200 m in thickness (Baudin and Berthou 1996), these layers are overlain by the 80–180 m thick Santana Formation. Showing interdigitation with the second cycle of the Batateira sandstones or foliated clays, the Crato member of this formation corresponds to 20–70 m of discontinuous laminated limestones interlayered with mudstones, siltstones, or shales (Martill et al. 2007; Fig. 2.9a). The overlying Ipubi Member, of Late Aptian age, is mainly composed of discontinuous and thick layers of gypsum (up to 20 m) alternating with black to brown or green foliated clays passing upwards to marls, clays, and sandstones (Romualdo Member, Early Albian) deposited after an unconformity indicating a short period of erosion (Assine 2007).

Abundant fossils in the Romualdo layers indicate a shallow marine ingression, coming from the SSE and having reached the west end of the basin. The overlying

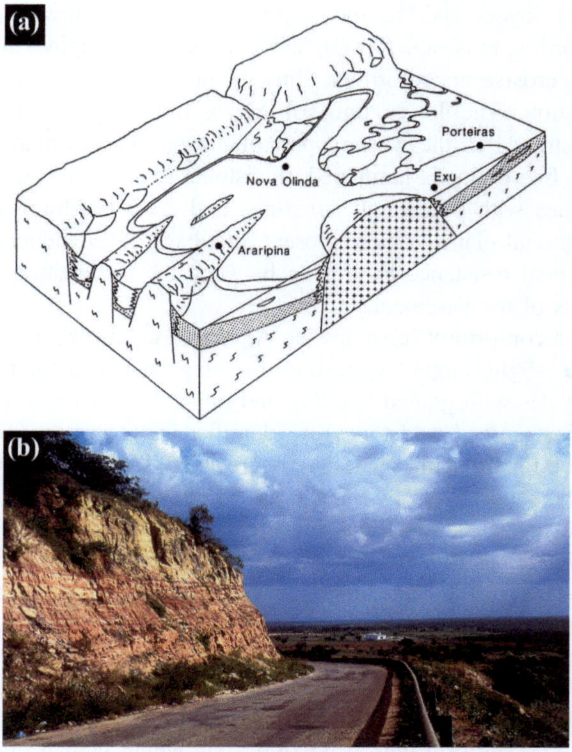

Fig. 2.9 Mesozoic post-rift sedimentary covers and their palaeogeographical implications. **a** Reconstruction of the lake/lagoon in which the Crato Member plattenkalks were deposited (Martill 1993); a large delta advances from the east and large fans of coarse clastic material are accumulating at the foot the escarpment produced by the Aurora fault zone (Nova Olinda). Presence of residual scarps and uncompletely beveled horsts and granitic hills (Exu). **b** Cuesta forming the south edge of the Chapada do Apodi, north of Apodi. Gradation between estuarine or deltaic sandstones of the upper part of the Açu Formation and the marine Jandaira limestone (Turonien-Campanien; *top*, *left*). Exhumed pre-Cenomanian surface truncating the basement at right and at the back. *Photograph* J.P. Peulvast

clayey sandstones are the last deposits of the lacustrine to marine phases of sedimentation. Reaching 20–80 m in thickness, they form the Arajara Formation of Ponte and Appi (1990), or the final part of the Santana Formation according to Assine (2007). Outliers of lacustrine limestones, similar to those of the Crato Member, exist in the Tucano basin, indicating a continuation of the lake systems to the south of the Pernambuco lineament. Other outliers containing the whole sequence are located in the Socorro basin (Serra Vermelha, Pernambuco-Piauí border) and the Serra Negra (Jatobá basin), at similar altitudes.

Whereas, the other basins of the eastern margin of northeast Brazil were still submitted to transgressive marine conditions until the Late Cretaceous, the end of the sedimentary sequence in the Araripe basin was continental. The alluvial sediments of the so-called Exu Formation (150–250 m in thickness) unconformably cover the older layers and, to the south, west, and northwest, the basement (Fig. 2.6). According to Assine (2007), they consist in two distinct lithological units separated by an erosive unconformity. Only the upper one should receive the name of Exu Formation. The lower unit, of Meso-Albian age, is called Araripina Formation. Restricted to the western part of the basin, it is made of rhythmites where reddish fine-grained, laminated sandstones contain plurimetric lenses of coarse sandstones with overweigh structures and cross-bedding, indicating sedimentation in alluvial plains. Where present, this basal layer induces a strong contrast in mechanical resistance to erosion between the sedimentary cover and the crystalline rocks of the basement.

An erosive unconformity separates the Araripina sediments from the Exu sandstones, with a slight angle suggesting ongoing synsedimentary deformation. Sedimentary cycles with graded bedding and upward decrease of grain size (conglomerates to fine-grained sandstones) are identified in the western (distal) part of the basin, where flood plain pelitic facies are intercalated. In the eastern (proximal) part of the basin, the sandstones are coarser and more immature, with abundant conglomeratic beds and cross beddings, reflecting sedimentation in braided channels. The age is considered as Albian-Cenomanian, and the paleoflows indicate a transport to the west (Assine 2007). The source area is the Borborema region whose uplift, more intense to the east, has begun as soon as the middle Albian (Assine 1994).

Whereas the Araripe post-rift cover presently culminates above 900 m, the post-rift sediments of the Potiguar basin form the low-lying Chapada do Apodi (80–150 m in altitude). The onshore basin corresponds to a broad (100 km wide or more) and shallow synform with a NE–SW axis which is also that of secondary compressive undulations deforming the post-rift cover (Maia and Bezerra 2014b). It is filled with a thick transgressive sequence whose lower layers, the 400–800-m thick Açu sandstones (Albian-Cenomanian) rest unconformably on syn-rift or transitional deposits and incompletely beveled basement highs (Sousa et al. 2008a), extending up to 30–40 km beyond the border fault of the buried rifts, on the planated Aracati Platform and Touros High (Castro et al. 1981; Araripe and Feijó 1994; see Fig. 3.3). Grading upwards to 100 m of fine estuarine or deltaic sediments (Souza 1982), these coarse fluvial sandstones are conformably covered by thick (up to 400 m) bioclastic calcarenites and calcilutites deposited together with minor siliciclastic sediments in estuaries, intertidal plains, beaches, lagoons, and shallow open sea of increasing depth from Turonian to Early Campanian (Jandaira Formation, 90–80 Myear; Córdoba and Castro 2002; Fig. 2.9b).

2.4.2 Offshore Basins and Formation of the Transform Margin

Marine conditions of sedimentation prevailed during the same period in the offshore Potiguar basin, which became part of the Equatorial continental margin after the Albian. The sandstones of the Açu Formation interdigitate with the more calcareous rocks of the Ponta do Mel Formation, and the limestones of the Jandaira Formation grade seaward and upward to the foliated clays, siltites, and sandstones of the regressive Ubarana Formation, deposited on the continental slope. After a transgressive phase (Albian-Early Campanian), the post-Campanian deposits make up a mixed platform where a large regressive cycle is recognized until the Holocene, forming a seaward thickening coastal-shelf-slope-basin system (Pessoa Neto 2003). These deposits are composed of three lithostratigraphic units known as the Tibau Formation (sandstones and conglomerates), the Guamaré Formation (limestones), and the Ubarana Formation (marine shales). These units respectively represent the siliciclastic sedimentation of the inner shelf, the carbonate deposits of the outer shelf, and the pelitic sedimentation on the continental slope. Several unconformities, stages of canyon incision and infilling (Neogene), and variations in the position of the limits between the clastic and carbonate deposition reflect eustatic fluctuations and variations in the clastic supply from the continent.

Deformations and unconformities have been identified in this series. After a long tectonically quiet period of subsidence, from Albian to Middle Campanian, the Potiguar basin underwent tectonic activity with the formation of the Ubarana transcurrent zone, 20-km north of the present coast line (Cremonini 1995, 1996; Cremonini et al. 1996). This event was followed by erosion—the Pre-Ubarana Unconformity—after which the offshore basin was covered by the Ubarana Formation. This unconformity is related to a major reorganization of the Potiguar Basin, with transient regional uplift, after which no sedimentation is known in the onshore part of the basin, until the deposition of Cenozoic clastic sediments in a narrow coastal zone (Bigarella 1975). This may have resulted from thermal uplift related to proximity of the mid-oceanic spreading zone, characterized by the presence of recently formed oceanic crust in front of the basin (Cremonini and Karner 1995), before this zone moved away along the recently created transform margin, resulting in later lithospheric cooling and subsidence (mechanism described by Boillot and Coulon 1998).

To the west of the Potiguar Basin and of the Caponga High, the Equatorial margin comprises several buried basins (the Mundaú, Icarai, Acaraú, and Piauí-Camocim sub-basins), separated by basement "highs" (Fortaleza and Ceará highs; Fig. 2.4). Forming the greater Ceará basin (34,000 km^2), these sub-basins lack the typical rift structures of most passive margins (half grabens, tilted blocks resulting from brittle extension and stretching of the crust: Boillot and Coulon 1998). According to Matos (2000, 2002), transform faulting linking the wide and shallow precursory basins began in the Albian, defining a lithospheric boundary between Africa and South America from the mouth of the Amazon to the Niger delta, without control by the basement structural fabric, and leading to the formation of oceanic crust. Wrench tectonics induced the formation of conjugate strike-slip fault systems linked to the newly formed oceanic transform fault zones (Chain, Romanche), with associated secondary normal faults, reverse faults and folded structures, respectively, in the releasing and restraining transform segments (Fig. 2.8).

The migration process of active transform faults resulted in diachronous relationships and significant differences between basins in terms of subsidence histories, geometries, and facies distribution. Whereas the East Potiguar (Benue) basin and Touros high evolved through transtension along a narrow zone, the Potiguar-Mundaú-Icarai (Keta-Togo-Benin) segment formed a broad region dominated by extension, limited by discrete shear zones, which evolved under pure shear-dominated transtension. To the west, a large transpressive belt developed in the Piauí-Camocim-Acaraú (Accra) basin, comprised of major oblique/strike-slip faults and *en échelon* folds. A typical passive margin environment was only reached later, with progressive oceanward migration of active transform zones, and after the thermal effect of nearby spreading centers was finished (Matos 2000, 2002).

2.5 Post-campanian Sedimentation and Volcanism: The "Drift Phase"

The increasing proportion of terrigenous facies among the deposits of the offshore Potiguar basin reflects an increase of onshore erosion since the Late Cretaceous. The total thickness of Cenozoic sediments on this part of the margin is locally important (up to 3200 m below the shelf break: Morais Neto 1999). A sharp increase in the clastic discharge recorded in the Miocene (especially Late Miocene) and younger series (Pessoa Neto 1999, 2003) was interpreted by this author and by Morais Neto et al. (2000) as a result of significant erosion processes—also reflected by the deposition of the Barreiras sediments—that would have affected the east Borborema Province in relation with uplift during Late Tertiary time, since 20 Myear.

In the Ceará basin, the continental input seems to have remained moderate and often discontinuous. Only Cretaceous deposits are found on the most salient areas. The diversity of the structural controls along the equatorial margin (Matos 2000), but also the strong asymmetry of the drainage patterns in the Jaguaribe-Piranhas hemicycle explain the heterogeneity of the sedimentation. Whereas, a long sedimentary hiatus of 50 Myear separates the Albian or Cenomanian and Late Eocene sequences in the Piauí-Camocim sub-basin (Beltrami et al. 1994), a less discontinuous sedimentation is recorded in the Acaraú and Icaraí sub-basins (shorter interruptions in the Cenomanian and Paleocene periods), and at least six erosive/non depositional events are recognized in the Mundaú sub-basin from Early Paleocene to Early Oligocene. The Oligocene period seems to have undergone higher rates of sedimentation, at least in the Mundaú sub-basin (Cunha 1991).

An important volcanism known offshore in the Tibau and Ubarana Formations (Araripe and Feijó 1994; Beltrami et al. 1994) of the Potiguar basin and of the Mundaú sub-basin (Macaú Formation: 31.1–22.6 Myear and 19.9–12.4 Myear: Vieira et al. 2002), and on the Fortaleza High (Mundaú Volcanism: 32 ± 1 Myear; Almeida et al. 1988; Fig. 2.2) has occurred at the continental termination of the volcanic ridge associated with the Chain Fracture Zone (Mizusaki et al. 2002). It is chronologically related to the onshore Oligocene Messejana Volcanism (30–34 Myear), mainly recognized around Fortaleza, with several necks and dykes of

Fig. 2.10 The «Serrote de Jericoacoara» (northwest Ceará), quartzitic residual hill protruding from the beveled basement through the Barreiras sediments and abruptly cut by the highest sea-cliff of the study area. The mega-barchans of Jericoacoara at the back. *Photograph* J.P. Peulvast

phonolite, trachyte, and basalt up to 35 km inland (Almeida et al. 1988; Carneiro et al. 1989). The Macaú volcanism is also found onshore, in the eastern Rio Grande do Norte and Paraiba states, forming scattered dykes and flows of olivine basalts, and plugs or necks of basanites and ankaratrites, in the basement (Pico de Cabugi) as well as in the Mesozoic cover of the Potiguar basin, within a 400-km long and 60 km wide strip trending NNW (Dantas et al. 2002). Most of them were intruded in the crust between 30 and 20 Myear (Silveira 2006), although ages from 57.0 to 6.4 Myear are reported (Oliveira and Medeiros 2012).

The onshore deposits differ from their offshore counterparts—the upper part of the Tibau Formation (Saadi and Torquato 1992; Araripe and Feijó 1994; Pessoa Neto 2003)—by several features such as limited thicknesses (less than 80 m and often less than 15 m) and relatively uniform distribution along the coast. Presenting many similarities with the "Continental Terminal" of West Africa (Lappartient 1985; Conrad and Lappartient 1987; Lang et al. 1990) or with the Cuddalore and Warkali beds of peninsular India (Gunnell 1996, 1998), these clastic sediments are known as part of the "Barreiras Group" (Mabesoone et al. 1972) identified along major parts of the northern and eastern Brazilian coasts (Bigarella 1975) and considered as mainly Miocene in age (Arai 2006; Lima 2008; Sousa et al. 2008a, b; Rossetti et al. 2011, 2013).

In the study area, the Barreiras Group sediments directly overlie the weathered basement (Cunha and Castelo Branco 2002), but also lie unconformably on Late Cretaceous sediments (Jandaira limestones) in the Potiguar Basin. They underlie the 5–50-km wide "tabuleiros", but narrow outcrops of the basement locally separate them from the continental shelf, forming weakly prominent points on the shoulder of the Messejana graben (Iguape) and on the Fortaleza High (Mucuripe, Pecém), and a well-marked salient at Jericoacoara (western Ceará), where high sea-cliffs cut the base of a residual quartzitic hill on the southern margin of the Acaraú sub-basin (Fig. 2.10). Covering the distal parts of the planated pedestal of the marginal

(a)
(b)
(c)
(d)
(e)
(f)

◀ **Fig. 2.11** The Barreiras Group sediments. **a** Deformed post-Turonian sediments and unconformable deposits (allochtonous laterite, paleodunes) at Ponta Peroba (Icapui). **b** Sea-cliffs at Morro Branco, exposing the upper facies of the Barreiras sediments, represented by yellowish, fine-grained sandstones of probable fluvio-marine origin, and covered at the top by oxidized, nonconsolidated palaeodune deposits. **c** Seismites in the Barreiras Group sediments: a sand dyke at Praia das Fontes (Beberibe). The Barreiras sediments are overlain by palaeodune material (at the back). **d** Quarry to the west of Beberibe, showing *red* to *yellow* layered argillaceous sands, containing minor amounts of quartz gravel and partly consolidated. **e** Camocim conglomerate (northwest Ceará): mature fluvial quartz conglomerate, with well-rounded pebbles cemented by a red silica-ferruginous matrix. **f** Laterite cap developed from iron-rich Barreiras sediments and fossilized by palaeodune material at Ponta Peroba (Icapui). *Photographs* J.P. Peulvast and F. Bétard

escarpment, they are reduced to erosive remnants disconnected from source areas and dissected by rivers flowing to the ocean (Bétard 2007). Therefore the "tabuleiros" form a succession of triangular units separated by outcrops of the substratum along the floors of the main valleys or by mountain ridges approaching the coast (Fortaleza High). They mostly overlook the sea by low active or abandoned sea-cliffs, 2–20-m high, resulting from marine erosion (Fig. 1.3).

The Barreiras sediments slightly dip seaward from a few tens of meters inland to sea level or lower at the coast. They present multiple vertical and lateral facies variations which have been studied in detail along the coast of the Potiguar basin, in eastern Ceará (Sousa et al. 2008a, b). Between Lagoa do Mato and Icapui, on a 40-km long sea-cliff tract (Fig. 2.11a), these sediments are found in horizontal layers, or highly deformed beds, mainly from Icapui to Retirinho. In the nondeformed sectors (e.g., between Redonda and Ponta Grossa), the cliffs expose the upper facies of the Barreiras sediments, represented by reddish medium- to coarse-grained sandstones with conglomerate intercalations and massive structure.

Although the lower level locally appears in this sector, directly overlying small outcrops of Jandaira limestone (W of Ponta Grossa), it rather forms the bulk of the deformed sectors, with fine- to medium-grained sandstones of yellow to purple or red colors indicating strong oxidation. Intercalations of layers of massive ferruginous sandstones or conglomerates with well-rounded to subangular quartz and crystalline pebbles are observed, often poorly sorted. Westward, these units pass laterally to yellowish-greenish, medium- to coarse-grained sandstones deprived of clear bedding, containing clay beds. Forming the lower parts of the sea-cliffs to the west, from Lagoa do Mato to Retirinho and beyond, they are also identified west of the Jaguaribe river, along the lower estuary and in Fortim, well beyond the limits of the Potiguar basin, as far as Beberibe (Praia das Fontes, Morro Branco; Fig. 2.11b), and even west of Fortaleza (Caucaia: Icarai). Many sea-cliff outcrops also expose coarser sediments. Ferruginous gravels and clay balls (bentonite of possibly volcanic origin, smectite), and also carbonate concretions and bioturbations (ichnofossils) are found in the conglomerates which often form the lower parts of the outcrops. Bedding may be discontinuous or lacking, obliterated by synsedimentary plastic deformations (Morro Branco, Praia das Fontes: Fig. 2.11c). These sediments are considered by Sousa et al. (2008a) as belonging to the Tibau Formation. A more or less developed lateritic horizon locally tops the sediments of both deformed and nondeformed units, below argillaceous sands which grade upwards to white, yellow to reddish sands with sparse subangular quartz gravels (Potengi Formation). Inland sections show layers of reddish to yellow or white argilaceous sands with minor amounts of quartz gravel (Fig. 2.11d), and more clay beds of variable thickness at depth (Cascavel: Cunha and Castelo Branco 2002).

Conglomeratic materials are best represented in the vicinity of present river mouths (Coreaú, Curu, Choró rivers) and of mountains approaching the coast (around the Baturité massif: Pecém-Taiba, Prainha). Around the Rio Coreaú mouth, the Camocim Formation forms a prominent sea-cliff where a mature fluvial quartz conglomerate, with well-rounded pebbles cemented by a red silica-ferruginous matrix, reaches a residual thickness of 20 m (Costa et al. 1979; Saadi and Torquato 1992; Fig. 2.11e). Thinner and often widespread conglomeratic formations are also identified around other river mouths (Jaguaribe).

The age of these non-fossiliferous sediments is now relatively well constrained on the basis of geochronological data obtained from K–Ar and $^{40}Ar/^{39}Ar$ dating of supergene minerals, and (U-Th)/He dating of supergene goethite grains taken in their weathering mantles (Lima 2008). According to these results, the maximum age of the Barreiras sediments in northeast Brazil would be 22 Myear (Aquitanian), their deposition would have continued until 16–17 Myear (Burdigalian), before the middle Miocene highstand, and their weathering until 7 Myear, indicating mainly warm and humid climatic conditions in the coastal area during this period. According to Rossetti et al. (2013), the prolonged sea-level lowstand recorded since the Tortonian may explain the prevalence of subaerial conditions and the lateritic soil formation from mid-Miocene up to late Quaternary.

The relationship with the overlying sediments (Potengi Formation) is not clear in all publications since many of them deal either with the northern or the eastern coasts of Northeast Brazil, which are not entirely comparable. In the east of the study area, these Post-Barreiras deposits (Rossetti et al. 2011) cover unconformably both the Barreiras and Tibau Formations and their laterite cap (Fig. 2.11f). Commonly referred to as palaeodune material (Brandão 1995; Sousa et al. 1999, 2008a), these more or less consolidated sand deposits of white to yellow or red colors are probably more complex, representing weathered material formed in dune, interdune (creeks and ponds), lagoon, and shore environments, as suggested by the presence of scattered coarse quartz grains (up to 1 cm) in the fine-grained material at all levels (Menezes et al. 1998; Claudino Sales 2002). Outcropping in the upper part of the sea-cliffs of eastern Ceará, they form the main part of most cliffs and paleo-cliffs west of Caucaia, except in a few rocky shore segments which expose the underlying Barreiras sediments or basement rocks (e.g., Taiba, Camocim, and Jericoacoara). More than the less extensive Barreiras sediments *s.s.*, they form the substrate of the so-called tabuleiros on large widths (Bétard 2007), below a discontinuous veil of unconsolidated white sand corresponding to various generations of Holocene to Mid-Pleistocene paleodunes (Barretto et al. 2004) (Fig. 2.12).

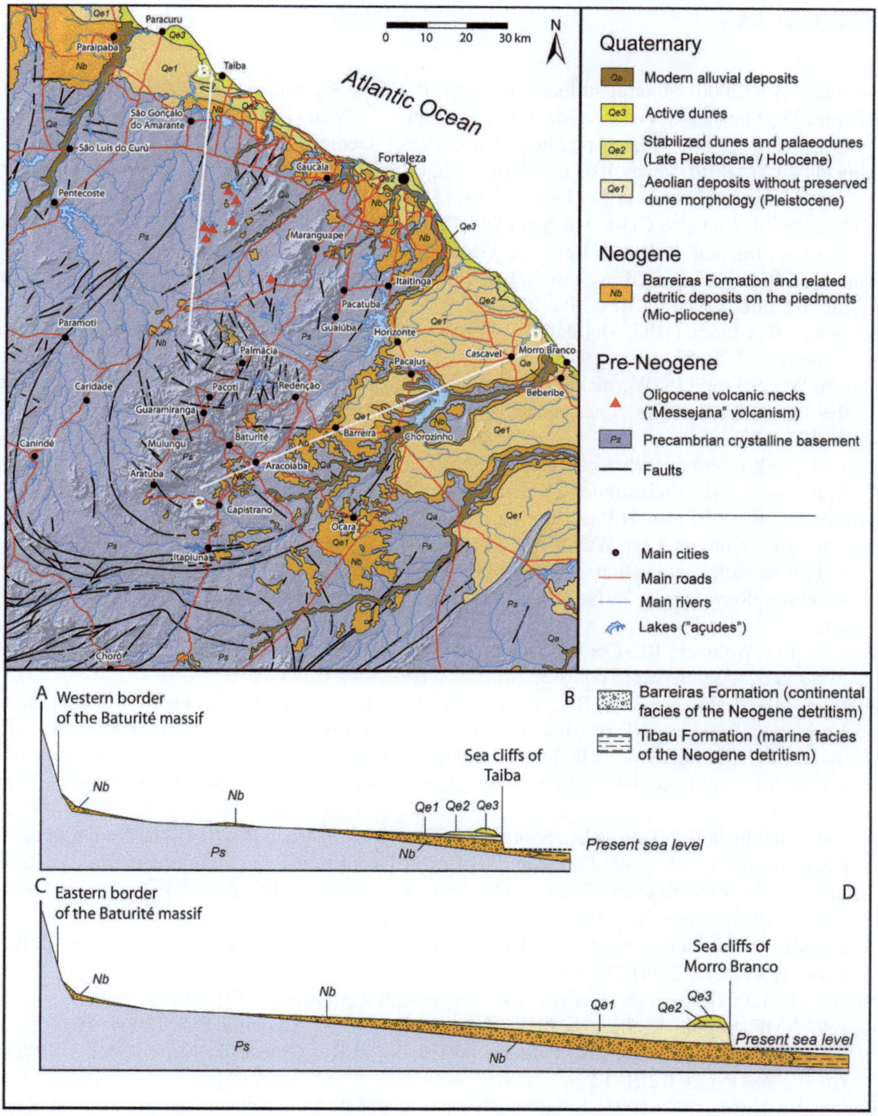

Fig. 2.12 Geological map of the Quaternary-Neogene deposits in the Baturité region, and synthetic profiles showing the geometric relationships between the different sets of deposits. From Bétard (2007)

References

Ab'Saber AN (2000) Summit surfaces in Brazil. Rev Bras Geociências 30(3):515–516

Alkmim FF, Marshak S, Fonseca MA (2001) Assembling West Gondwana in the Neoproterozoic: clues from the São Francisco craton region, Brazil. Geology 29:319–322

Almeida FFM, Brito Neves BB, Carneiro CDR (2000) The origin and evolution of the South American platform. Earth Sci Rev 50:77–111

Almeida FFM, Carneiro CDR, Machado DL, Dehira LK (1988) Magmatismo pós-paleozóico no Nordeste oriental do Brasil. Rev Bras Geociências 18(4):451–462

Almeida FFM, Hasui Y, Brito Neves BB, Fuck RA (1981) Brazilian structural provinces: an introduction. Earth Sci Rev 17:1–21

Almeida AR, Ulbrich HHG, McReath I (1999) O batólito Quixadá – petrologia e geoquímica. Rev Geologia 12:29–52

Amaral WS, Santos TJS, Wernick E (2011) Occurrence and geochemistry of metamafic rocks from the Forquilha Eclogite zone, central Ceará (NE Brazil): geodynamic implications. Geol J 46:137–155

Arai M (2006) Agrande elevação eustática do Mioceno e sua influência na origem do Grupo Barreiras. Revista do Instituto de Geociências, USP. Geol. USP Sér. Cient., São Paulo 6(2):1–6

Araripe PT, Feijo FJ (1994) Bacia Potiguar. Bol Geoci Petrobras 8(1):127–141

Araújo CEG, Cordani UG, Weinberg RF, Basei MAS, Armstrong R, Sato K (2014) Tracing Neoproterozoic subduction in the Borborema Province (NE Brazil): clues from U–Pb geochronology and Sr–Nd–Hf–O isotopes on granitoids and migmatites. Lithos 202–203:167–189

Araújo CEG, Weinberg RF, Cordani UG (2013) Extruding the Borborema Province (NE Brazil): a two-stage Neoproterozoic collision process. Terra Nova 0:1–12

Arthaud MH, Caby R, Fuck RA, Dantas EL, Parente CV (2008) Geology of the northern Borborema Province, NE Brazil and its correlation with Nigeria, NW Africa. In: Pankhurst RJ, Trouw RAJ, Brito Neves BB, De Wit MJ (eds) West Gondwana Pre-Cenozoic correlations across the South Atlantic region. Geological Society, London (Special Publications 294), pp 49–67

Arthaud MH, Fuck RA, Dantas EL, Santos TJS, Caby R, Armstrong R (2014) The Neoproterozoic Ceará Group, Ceará Central domain, NE Brazil: Depositional age and provenance of detrital material. New insights from UePb and SmeNd geochronology. J South Am Earth Sci 56:54–67. doi:10.1016/j.jsames.2014.06.011

Assine ML (1994) Paleocorrentes e paleogeografia na Bacia do Araripe, Nordeste do Brasil. Rev Bras Geociências 24(4):223–232

Assine ML (2007) Bacia do Araripe. Bol. Geociências Petrobrás 15:371–389

Barreto AMF, Suguio K, Bezerra FHR, Tatumi SH, Yee M, Giannini PCF (2004) Geologia e geomorfologia do Quaternário Costeiro do Estado do Rio Grande do Norte. Geol. USP Sér. Cient., São Paulo 4(2):1–12

Baudin F, Berthou PY (1996) Environnement de dépôt de la matière organique des sédiments aptiens-albiens du basin d'Araripe (NE du Brésil). Bull Centres Rech Explor Prod Elf-Aquitaine 20(1):213–227

Beltrami CV, Alves LEM, Feijo F (1994) Bacia do Ceará. Bol Geoci Petrobras 8(1):117–125

Benkhelil J (1988) Structure et évolution géodynamique du bassin intracontinental de la Bénoué (Nigéria). Bull Centres Rech Explor Prod Elf-Aquitaine 12(1):29–128

Bertani RT, Costa IG, Matos RMD (1990) Evolução tectono-sedimentar, estilo estrutural e habitat do petróleo na Bacia Potiguar. In: Raja Gabaglia GP, Milani EJ (eds) Origem e evolução de bacias sedimentares. Petrobras, pp 291–309

Bétard F (2007) Montagnes humides au cœur du Nordeste brésilien semi-aride: le cas du massif de Baturité (Ceará). Apports d'une approche morphopédologique pour la connaissance et la gestion des milieux. Ph.D thesis, Univ Paris-Sorbonne

Bigarella JJ (1975) The Barreiras Group in Northeastern Brazil. An Acad Bras Ciên 47:365–393

Boillot G, Coulon C (1998) La déchirure continentale et l'ouverture océanique. Géologie des marges passives. Gordon and Breach Science Publishers, Amsterdam

Brandão RL (1995) Mapa geológico da região metropolitana de Fortaleza. Texto explicativo. Sistema de informações para gestão et administração territorial da Região metropolitana de Fortaleza – Projeto SINFOR, CPRM

Brito Neves BB, Dos Santos EJ, Van Schmus WR (2000) Tectonic history of the Borborema Province, northeastern Brazil. In: Cordani U, Milani EJ, Thomaz Filho A, Campos DA (eds) Tectonic evolution of South America. 31st International Geological Congress, Rio de Janeiro, Brazil, pp 151–182

Caby R, Arthaud M (1986) Major Precambrian nappes of the Brazilian Belt, Ceará, northeast Brazil. Geology 14:871–874

Caby R, Arthaud MH, Archanjo CJ (1995) Lithostratigraphy and petrostructural characterization of supracrustal units in the Brasiliano Belt of Northeast Brazil: geodynamic implications. J South Am Earth Sci 8(3–4):235–246

Carneiro CDR, Hamza VM, Almeida FFM (1989) Ativação tectônica, fluxo geotérmico e sismicidade no Nordeste oriental brasileiro. Rev Bras Geociências 19(3):310–322

Carneiro CDR, Monma R, Ferreira FJF, Dehira LK, Magalhães FS (1988) Ensaio de integração geológico-geofísica sobre o Nordeste oriental com base em mapas aeromagnéticos. XXXV Congr Bras Geol, Belém, Pará, Anais 5:2122–2136

Castro JC, Barrocas SLS, Lima HP (1981) Fácies, diagênese e modelos de acumulação da Formação Açu, parte emersa da Bacia Potiguar. Cenpes, Petrobras

Claudino Sales V (2002) Evolution géomorphologique de la zone côtière de Ceará (Nord-est du Brésil): du long terme au court terme. PhD thesis, Univ Paris-Sorbonne

Coimbra JC, Arai M, Carreño AL (2002) Biostratigraphy of Lower Cretaceous microfossils from the Araripe Basin, northeastern Brazil. Geobios 35:687–698

Conrad G, Lappartient JR (1987) Le 'Continental Terminal', sa place dans l'évolution géodynamique du bassin sénégalo-mauritanien durant le Cénozoïque. J Afr Earth Sci 6:45–60

Córdoba VC, Castro JC (2002) Analíse paleoambiental e diagenética da rampa carbonática Jandaíra, Neocretáceo na Bacia Potiguar. XLI Congr Bras Geol, João Pessoa, Anais, p 264

Corsini M, Vauchez A, Archanjo CJ, Jardim de Sá EF (1991) Strain transfer at continental scale from a transcurrent shear zone to a transpressional fold belt: the Patos-Seridó system, northeastern Brazil. Geology 19:586–589

Corsino AR, Tiriba VF (1985) Nota sobre a occorrência de uma pequena bacia sedimentar ao norte do município de Umarizal (R.N) – (Bacia de Gangorra). Bol DG-CCE/UFRN, Natal 9:54–56

Costa MJ, Lins CAC, Bacchiegga JF, Habekost CR, Cruz B (1979) Geologia da Bacia Jaibaras, Ceará, Piauí e Maranhão (Projeto Jaibaras). MME, Geologia Básica 11

Costa IG, Beltrami CV, Alves LEM (1990) A evolução tectono-sedimentar e o habitat do óleo da Bacia do Ceará. Bol Geoci Petrobras 4(1):65–74

Costa ACD, Hackspacher PC, Dantas EL, Fetter AH (2002) A zona de cisalhamento Patos ou "sutura tectônica Patos» no Domínio da Zona Transversal, Provincia Borborema: seu significado na colagem neoproterozóica. XLI Congr Bras Geol, João Pessoa, Anais, p 300

Cremonini OA (1995) A reativação tectônica da Bacia Potiguar no Cretáceo superior. V Simp Nac Estudos Tectônicos, Gramado-RS, Anais, pp 277–280

Cremonini OA (1996) Evolução tectônica da area de Ubarana, porção submersa da Bacia Potiguar. Bol Geoci Petrobras 10(1–4):81–97

Cremonini OA, Karner GD (1995) Soerguimento termal e erosão na Bacia Potiguar submersa, e seu relacionamento com a evolução da Margem Equatorial brasileira. XVI Simp. Geol. do Nordeste, Recife

Cremonini OA, Goulart JPM, Soares UM (1996) O rifte Potiguar: novos dados e implicações tectonicas. Bol 4o Simp sobre o Cretáceo do Brasil, UNESP, pp 89–93

Cunha AAS (1991) Reconhecimento dos hiatos da seção terciária da sub-bacia de Mundaú (Bacia do Ceará): um estudo baseado em nanofosseís calcários. Bol Geoci Petrobras 8(1):69–82

Cunha LS, Castelo Branco RMG (2002) Configuração 3-D dos sedimentos quaternaries baseada em inverses de dados geofísicos elétricos e eletromagnéticos, região costeira de Cascavel – Ceará. XLI Congr Bras Geol, João Pessoa, Anais, p 85

Dantas AR, Almeida RB, Sena ES, Bezerra FHR, Oliveira DC (2002) Uso de SIG no estudo da distribuição do evento ígneo Macau, Nordeste oriental, Brasil. XLI Congr Bras Geol, João Pessoa, Anais, p 157

De Castro DL, Medeiros WE (1997) Afinamento crustal associado à implantação das bacias intracratônicas do trend Cariri-Potiguar. XVII Simp Geol do Nordeste, Fortaleza, Anais, pp 335–338

De Castro DL, Castelo Branco RMG (1999) Caracterização da arquitetura interna da bacias do Vale do Cariri (NE do Brasil) com base em modelagem gravimétrica 3-D. Braz J Geophys 17:129–144

De Castro DL, Costa FCN (1999) Mapas gravimétricos do Estado do Ceará. Rev Geologia 12:69–80

De Castro DL, Castelo Branco RMG, Martins G, Castro NA (2002) Radiometric, magnetic, and gravity study of the Quixadá batholith, central Ceará domain (NE Brazil): evidence for Pan-African/Brasiliano extension-controlled emplacement. J South Am Earth Sci 15:543–551

De Castro DL, Oliveira DC, Castelo Branco RMG (2007) On the tectonics of the Neocomian Rio do Peixe rift basin, NE Brazil: lessons from gravity, magnetic and radiometric data. J South Am Earth Sci 24:184–202

De Castro DL, Bezerra FHR, Castelo Branco RMG (2008) Geophysical evidence of crustal-heterogeneity control of fault growth in the Neocomian Iguatu basin, NE Brazil. J South Am Earth Sci 26:271–285

De Castro DL, Bezerra FHR, Sousa MOL, Fuck RA (2012) Influence of Neoproterozoic tectonic fabric on the origin of the Potiguar Basin, northeastern Brazil and its link with West Africa based on gravity and magnetic data. J Geodyn 54:29–42

De Castro DL, Fuck RA, Phillips JD, Vidotti RM, Bezerra FH, Dantas EL (2014) Crustal structure beneath the Paleozoic Parnaíba Basin revealed by airborne gravity and magnetic data, Brazil. Tectonophysics 614:128–145

De Castro DL, Barbosa VCF, Silva JBC, Medeiros WE (1997) Relevo da interface crosta-manto entre os vínculos de isostasia e suavidade. Congr. Int. SBGf 5, São Paulo, Anais, pp 682–685

Destro N, Szatmari P, Ladeira EA (1994) Post-Devonian transpressional reactivation of a Proterozoic ductile shear zone in Ceará, NE Brazil. J Struct Geol 16(1):35–45

Ferreira JM, Oliveira R, Takeya MK, Assumpção MA (1998a) Superposition of local and regional stresses in NE Brazil: evidence from mechanisms around the Potiguar marginal basin. Geophys J Int 134:341–355

Ferreira VP, Sial AN, Jardim de Sa EF (1998b) Geochemical and isotopic signatures of Proterozoic granitoids in terranes of the Borborema structural province, northeastern Brazil. J South Am Earth Sci 11(5):439–455

Fetter A, Van Schmus WR, Dos Santos TJS, Arthaud M, Nogueira Neto J (1999) Geocronologia e estrutura do Estado do Ceará : NW da província Borborema, NE Brasil. XVII Simp Geol do Nordeste, Fortaleza, Anais, pp 32–33

Fetter AH, Van Schmus WR, Dos Santos TJS, Arthaud M, Nogueira Neto JA (2000) U-Pb and Sm-Nd geochronological constraints on the crustal evolution and basement architecture of Ceará State, NW Borborema Province, NE Brazil: implications for the existence of the Paleoproterozoic supercontinent Atlantica. Revista Brasileira de Geociências 30:102–106

Fetter AH, Dos Santos TS, Van Schmus WR, Hackspacher PC, De Brito Neves BB, Arthaud MH, Wernick E (2003) Evidence for Neoproterozoic continental arc magmatism in the Santa Quitéria Batholith of Ceará State, NW Borborema Province, NE Brazil: implications for the assembly of west Gondwana. Gondwana Res 6(2):265–273

Françolin JBL, Szatmari P (1987) Mecanismo de rifteamento da porção oriental da margem norte brasileira. Revista Brasileira de Geociências 17(2):196–207

Françolin JBL, Cobbold PR, Szatmari P (1994) Faulting in the Early Cretaceous Rio do Peixe basin (NE Brazil) and its significance for the opening of the Atlantic. J Struct Geol 16(5):647–661

Galvão CC, Da Silva FCA (2002) Análise microestrutural de um corpo siliciclástico localizado a Nordeste de Santana do Acaraú – CE. XLI Congr Bras Geol, João Pessoa, Anais, p 624

Godard A, Lagasquie JJ, Lageat Y (2001) Basement regions. Springer, Berlin

Guiraud R, Bellion Y (1995) Late Carboniferous to Recent geodynamic evolution of the West Gondwanian, cratonic, Tethyan margins. In: Nairn AEM et al (eds) The ocean basins and margins, vol 8., The Tethys oceanPlenum Press, New York, pp 101–124

Gunnell Y (1996) Géodynamique d'une moyenne montagne tropicale. La genèse des paysages dans le Ghat occidental du Deccan, sur son revers continental et son piémont maritime. PhD thesis, Univ. B. Pascal, Clermont-Ferrand

Gunnell Y (1998) Passive margin uplifts and their influence on climatic change and weathering patterns of tropical shield regions. Global Planet Change 18:47–57

Hegarty KA, Morais Neto JM, Karner GD (2002) The enigma of the Araripe plateau: new constraints on its uplift and tectonic history using AFTA. XLI Congr Bras Geol, Joã Pessoa, Anais, p 495

Hollanda MHBM, Pimentel MM, Oliveira DC, Jardim de Sá EF (2006) Lithosphere–asthenosphere interaction and the origin of Cretaceous tholeiitic magmatism in Northeastern Brazil: Sr–Nd–Pb isotopic evidence. Lithos 86:34–49

Jardim de Sá EF, Matos RMD, Morais Neto JM, Saadi A, Pessoa Neto OC (1999) Epirogenia Cenozózoica na Província Borborema : síntese e discussão sobre os modelos de deformação associados. VII Simp. Nac. Estudos Tect., Lençoís – Bahia, Anais, pp 58–61

Koutsoukos EAM (1992) Late aptian to maastrichtian foraminiferal biogeography and palaeoceanography of the Sergipe basin, Brazil. Palaeogeogr Palaeoclim Palaeoecol 92:295–324. doi:10.1016/0031-0182(92)90089-N

Lang J, Kogbe C, Alidou S, Alzouma KA, Bellion G, Dubois D, Durand A, Guiraud R, Houessou A, De Klasz I, Romann E, Salard-Cheboldaeff M, Trichet J (1990) The continental terminal in West Africa. J Afr Earth Sci 10:79–99

Lappartient JR (1985) Le Continental Terminal et le Pléistocène ancien du bassin sénégalo-mauritanien: stratigraphie, sédimentation, diagenèse, altération; reconstitution des paléorivages au travers des cuirasses. PhD thesis, Univ Aix-Marseille 3

Lima MG (2008) A História do Intemperismo na Província Borborema Oriental, Nordeste do Brasil. Implicações paleoclimáticas e tectônicas. PhD thesis, Universidade Federal do Rio Grande do Norte, Natal

Mabesoone JM, Campos e Silva A, Beurlen K (1972) Estratigrafia e origem do Grupo Barreiras em Pernambuco, Paraíba e Rio Grande do Norte. Rev Bras Geoc 2:173–188

Magnavita LP, Davison I, Kuznir NJ (1994) Rifting, erosion and uplift history of the Reconcavo-Tucano-Jatoba Rift, northeast Brazil. Tectonics 13:367–388

Maia RP, Bezerra FHR (2014a) Tópicos de geomorfologia estrutural. Nordeste brasileiro, Edições UFC, Fortaleza

Maia RP, Bezerra FHR (2014b) Inversão neotectônica do relevo na Bacia Potiguar, Nordeste do Brasil. Revista Brasileira de Geomorfologia 15(1):61–74

Martill DM (1993) Fossils of the Santana and Crato formations, Brazil (Field Guide to Fossils no. 5). The Palaeontological Association, London

Martill DM, Bechly G, Loveridge RF (2007) The Crato fossil beds of Brazil: window into an ancient world. Cambridge University Press, Cambridge

Matos RMD (1992) The Northeast Brazilian rift system. Tectonics 11(4):766–791

Matos RMD (2000) Tectonic evolution of the Equatorial South Atlantic. In: Mohriak W, Taiwani M (eds) Atlantic rifts and continental margins. Geophys Monograph 115, American Geophysical Union, pp 331–354

Matos RMD (2002) Geologia regional da margem equatorial afro-brasileira: evolução cinemática e o contexto paleogeográfico. XLI Congr Bras Geol, João Pessoa, Anais, p 272

Maurin JC, Guiraud R (1993) Basement control in the development of the early Cretaceous West and Central African rift system. Tectonophysics 228:81–95

Mello UT (1989) Controles tectônicos na estratigrafia da Bacia Potiguar: uma integração de modelos geodinâmicos. Bol Geoci Petrobras 3(4):347–364

Menezes MRF, Dantas EP, Caldas LH (1998) Estudos sedimentológicos na formação Tibau, porção emersa da Bacia Potiguar. Congr Bras Geol, Belo Horizonte, Anais, p 77

Menezes MRF, Lima Filho FP (1997) Estudo faciológico na Formação Serra do Martins, Serra de Portalegre - RN. XVII Simp Geol do Nordeste, Fortaleza 15:193–198

Menezes MRF, Lima Filho FP, Farias PRC (2002) Fácies e sistema deposicional da Formação Serra do Martins nos platôs terciários do Rio Grande do Norte. XLI Congr Bras Geol, Joã Pessoa, Anais, p 387

Mizusaki AMP, Thomaz-Filho A, Milani EJ, De Cesero P (2002) Mesozoic and Cenozoic igneous activity and its tectonic control in northeastern Brazil. J South Am Earth Sci 15:183–198

Monié P, Caby R, Arthaud M (1997) The Neoproterozoic Brasiliano orogeny in northeast Brazil: $^{40}Ar/^{39}Ar$ and petrostructural data from Ceará. Precambrian Res 81:241–264

Morais Neto JM (1999) As coberturas sedimentares tertiarias do interior da Paraiba e do Rio Grande do Norte e gênese da anteclise da Borborema. MSc thesis, Univ Federal de Ouro Preto, Ouro Preto, Brazil

Morais Neto JM, Alkmim FF (2001) A deformação das coberturas terciárias do Planalto da Borborema (PB-RN) e seu significado tectônico. Rev Bras Geociências 31(1):95–106

Morais Neto JM, Green PF, Karner GD, Alkmim FF (2008) Age of the Serra do Martins Formation, Borborema Plateau, northeastern Brazil: constraints from apatite and zircon fission track analysis. Bol Geoci Petrobras 16(1):23–52

Morais Neto JM, Hegarty KA, Karner GD, Alkmim FF, Matos RMD (2000) Uplift and erosion in the Borborema Province, northeast Brazil: insights from apatite fission track analysis. XXXI Int Geol Congr, IUGS/SBG, Abstract Volume, Rio de Janeiro

Morais Neto JM, Menezes MRF, Szatmari P, York D (2002) Datação Ar/Ar do plug basáltico «Serrote Preto» e seu significado para a cronologia de Formação Serra do Martins. XLI Congr Bras Geol, João Pessoa, Anais, p 499

Neumann VH (1999) Estratigrafía, sedimentología, geoquímica y diagénesis de los sistemas lacustres aptiense-albienses de la Cuenca de Araripe (Noreste de Brasil). PhD thesis, Univ Barcelona

Neumann VH, Borrego AG, Cabrera L, Dino R (2003) Organic matter composition and distribution through the Aptian-Albian lacustrine sequences of the Araripe basin, northeastern Brazil. Int J Coal Geol 54:21–40

Neves CAO (1989) Analise regional do trinomio geração-migração-accumulação de hidrocarbonetos na seqüência continental eocretácica da Bacia Potiguar. Bol Geoci Petrobras 3(3):131–145

Neves SP (2002) Considerações sobre a evolução termomecânica das faixas Orós, Piancó-Alto Brígida e Seridó, Província Borborema, Nordeste do Brasil. XLI Congr Bras Geol, João Pessoa, Anais, p 322

Neves SP (2015) Constraints from zircon geochronology on the tectonic evolution of the Borborema Province (NE Brazil): Widespread intracontinental Neoproterozoic reworking of a Paleoproterozoic accretionary orogen. J S Am Earth Sci 58:150–164. doi:10.1016/j.jsames.2014.08.004

Nogueira JF, Morales N, Nogueira Neto JA (1997) A estrutura interna do batólito de Quixadá – CE: dados preliminares. XVII Simp Geol Nordeste, Fortaleza, Anais, pp 297–301

Nogueira FCC, Marques FO, Bezerra FHR, De Castro DL, Fuck RA (2015) Cretaceous intracontinental rifting and post-rift inversion in NE Brazil: insights from the Rio do Peixe basin. Tectonophysics 644–645:92–107. doi:10.1016/j.tecto.2014.12.016

Oliveira DC, Gomes CJS (1996) A expressão de reativações pós-intrusão do enxame de diques Rio Ceará Mirim (Mesozóico): implicações na evolução tectônica da Bacia Potiguar. Bol Geoci Petrobras 10(1–4):121–145

Oliveira CG, Maas MVR (2002) Modelo descritivo das brechas quartzo-hematíticas do sudoeste do Cinturão cuprífero Orós-Jaguaribe, Província Borborema. XLI Congr. Bras. Geol., João Pessoa, Anais, p 459

Oliveira DC, Mohriak WU (2003) Jaibaras trough: an important element in the early tectonic evolution of the Parnaíba interior sag basin, Northern Brazil. Mar Pet Geol 20:351–383

Oliveira RG, Medeiros WE (2012) Evidences of buried loads in the base of the crust of Borborema Plateau (NE Brazil) from Bouguer admittance estimates. J S Am Earth Sci 37:60–76

Pessoa Neto OC (1999) Análise estratigráfica integrada da plataforma mista (siliciclástica-carbonática) do Neogeno da Bacia Potiguar, nordeste do Brasil. MSc thesis, Univ Fed Rio Grande do Sul, Porto Alegre, Brazil

Pessoa Neto OC (2003) Estratigrafia de seqüências da plataforma mista neogênica na Bacia Potiguar, margem equatorial brasileira. Rev Bras Geociências 33:263–278

Peulvast JP, Bétard F (2015) A history of basin inversion, scarp retreat and shallow denudation: the Araripe Basin as a keystone for understanding long-term landscape evolution in NE Brazil. Geomorphology 233:20–40. doi:10.1016/j.geomorph.2014.10.009

Peulvast JP, Claudino Sales V, Bétard F, Gunnell Y (2008) Low post-Cenomanian denudation depths across the Brazilian Northeast: implications for long-term landscape evolution at a transform continental margin. Global Planet Change 62:39–60. doi:10.1016/j.gloplacha.2007.11.005

Ponte FC, Appi CJ (1990) Proposta de revisão da coluna litoestratigráfica da Bacia do Araripe. In Congresso Brasileiro de Geologia 36, Anais, Natal, SBG, pp 211–226

Ponte FC, Ponte-Filho FC (1996) Estrutura geológica e evolução tectônica da Bacia do Araripe. MME/DNPM, Recife

Rossetti DF, Bezerra FHR, Góes AM, Brito Neves BB (2011) Sediment deformation in Miocene and Post-Miocene strata, Northeastern Brazil: evidence for paleoseismicity in a passive margin. Sed Geol 235:172–187

Rossetti DF, Bezerra FHR, Dominguez JLR (2013) Late Oligocene-Miocene transgressions along the equatorial and eastern margins of Brazil. Earth-Sci Rev 123:87–112

Saadi A, Torquato JR (1992) Contribuição à neotectônica do Estado do Ceará. Rev Geologia 5: 5–38

Santos TJS, Fetter AH, Hackspacher PC, Van Schmus WR, Neto JN (2008) Neoproterozoic tectonic and magmatic episodes in the NW sector of Borborema Province, NE Brazil, during assembly of Western Gondwana. J S Am Earth Sci 25(3):271–284

Sénant J, Popoff M (1991) Early Cretaceous extension in northeast Brazil related to the South Atlantic opening. Tectonophysics 198:35–46

Sial AN, Long LE, Pessoa DAR, Kawashita K (1981) Potassium-argon ages and strontium isotope geochemistry of Mesozoic and Tertiary basaltic rocks, northeastern Brazil. Acad. Brasileira de Ciências, 53, 115–122

Sial AN, Ferreira VP, Almeida AR (2002) C- and O isotope fluctuations in carbonates of the Freicheirinha Formation, NW Ceará, northeast Brazil. XLI Congr Bras Geol, João Pessoa, Anais, p 508

Silveira FV (2006) Magmatismo cenozóico da porção central do Rio Grande do Norte, NE do Brasil. Dissertation, Universidade Federal do Rio Grande do Norte, Natal

Sousa DC, Jardim De Sá EF, Matos RMD, Oliveira DC (1999) Deformação sin- e pós-Formação Barreiras na região de Ponta Grossa (Icapuí, CE), litoral ocidental da Bacia Potiguar. VII Simp Nac Estudos Tect, Lençoís – Bahia, Anais, pp 90–93

Sousa DC, Jardim de Sá EF, Antunes AF (2008a) Deformação neógena e suas implicações na estruturação dos campos de petróleo na região de Icapuí-Ponta Grossa (CE), Bacia Potiguar emersa. Revista Brasileira de Geologia 38(2-suplemento):97–110

Sousa DC Jardim de Sá EF, Vital H, do Nascimento MAL (2008b) Falésias na praia de Ponta Grossa, Icapuí, CE. Importantes deformações tectônicas cenozóicas em rochas sedimentares da Formação Barreiras. In: Winge M, Schobbenhaus C, Souza CRG, Fernandes ACS, Berbert-Born M, Queiroz ET (eds) Sítios geológicos e paleontológicos do Brasil. Published in Internet 24/09/2008. http://www.und.br/ig/sigep/sitio120.pdf

Souza SM (1982) Atualização da litoestratigrafia da Bacia Potiguar. XXXII Congr Bras Geol 32, Salvador, Anais 5, pp 2392–2406

Srivastava HK, Lima MS, Nogueira AM, Martins G, Pereira MA (1989) Geologia e estratigrafia das bacias de Pau dos Ferros e do Rio Nazaré (Rio Grande do Norte). XII Simp Geologia do Nordeste, Fortaleza, Anais, pp 157–161

Teixeira AL, Gaucher C, Paim PSG, Fonseca MD, Parente CV, Silva Filho WD, Almeida AD (2004) Bacias do estágio da transição da Plataforma Sul-Americana. In: Mantesso Neto V, Bartorelli A, Carneiro CDR, Brito-Neves BB (eds) Geologia do Continente Sul-Americano: Evolução da Obra de Fernando Flávio Marques de Almeida, Beca, pp 487–537

Torquato JR (ed) (1989) Granitóides do Ceará. Região de Quixadá-Solonópole. Rev Geologia 2 (1–2):1–143

Trompette R (1994) Geology of Western Gondwana (2000–500 Ma). Pan-African-Brasiliano aggregation of South America and Africa. Balkema, Rotterdam

Valença LMM, Neumann VH, Mabesoone JM (2003) An overview on Callovian-Cenomanian intracratonic basins of Northeast Brazil: onshore stratigraphic record of the opening of the southern Atlantic. Geol Acta 1(3):261–275

Vauchez A, Neves S, Caby R, Corsini M, Egydio-Silva M, Arthaud M, Amaro V (1995) The Borborema shear zone system, NE Brazil. J S Am Earth Sci 8(3–4):247–266

Vieira TM, Martins G, Oliveira DC, De Castro DL, Castelo Branco RMG (2002) O magmatismo alcalino de idade terciária das bacias Potiguar e Ceará: uma revisão comparativa preliminar. XLI Congr Bras Geol, João Pessoa, Anais, p 399

Chapter 3
Morphostructural Patterns: Influence of an Aborted Rift Zone on the Regional Organization of Landforms

Abstract A wide half-circle of crystalline and sedimentary highlands surrounds the "Sertaneja depression," opening onto the Equatorial Atlantic transform margin. Intersected seaward by offset elements of a 600 km long escarpment which overlooks coastal lowlands, it contains a roughly concentric system of erosion plains and scattered mountains shaped into a Precambrian basement whose geology is organized around a large transcurrent shear zone system of the Brasiliano orogen. The low surfaces slope gently seaward, disappearing below a coastal strip of Cretaceous and/or Tertiary sediments. The distribution of marginal highlands is related to the intersection of a failed rift zone by the continental margin. The discontinuous outlines of the marginal escarpment and its morphology strongly depend on the structures of the basement and of the aborted rift zone. The analysis of drainage patterns and of relationships between tectonic structures and landforms created by differential erosion allow an evaluation of the determining influence of superposed structural heritages in the present landscapes. Among these structural heritages, the numerous shear zones formed at depth in ductile regime during the Brasiliano orogeny play one of the most important roles, as well by focusing the later brittle deformations when conveniently oriented with regard to the stress fields, as by forming weakness zones when exposed to erosion.

Keywords Crustal thinning zone · Differential erosion · Epigeny · Fault scarp · Cuesta · Glint · Inselberg · Pediment · Rift shoulder · Structural landforms

3.1 The Main Units

The morphostructural patterns of the study area do not fully reflect the tectonic framework, since features of topographic inversion coexist with landforms more or less in conformity with this framework. For example, the highland rings comprise

J.-P. Peulvast and F. Bétard, *Landforms and Landscape Evolution of the Equatorial Margin of Northeast Brazil*, Springer Earth System Sciences,
DOI 10.1007/978-3-319-18203-2_3

parts of the Cariri-Potiguar rift basins (Araripe) as well as parts of their shoulders (Central Ceará highlands, Borborema), whereas interior lowlands also extend both on rift basins and faulted basement (Fig. 3.1). These discrepancies reflect various types of structural controls and show interferences between tectonic and erosional histories.

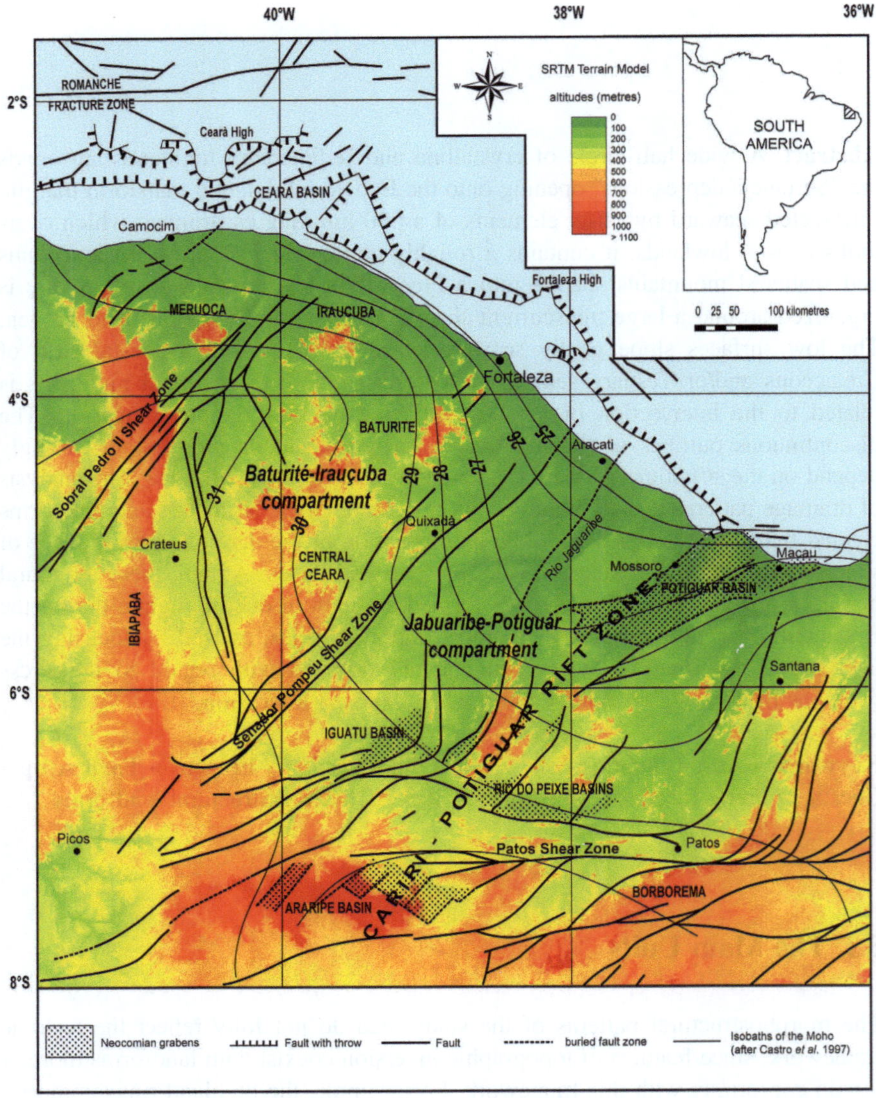

Fig. 3.1 Topography, rift structures and their links with crustal thickness in the equatorial margin of Northeast Brazil. Topography derived from the SRTM DEM

3.1.1 Structural Landforms of the Cariri-Potiguar Rift Zone

The Cariri-Potiguar rift zone is far from forming a simple geomorphological unit, because of its structural complexity and of the contrasting post-rift histories of its opposite extremities, which induced various degrees and natures of erosive shaping (Figs. 3.2 and 3.3).

3.1.1.1 Contrasted Landscapes of "Chapadas"

Both extremities of the rift zone form plateaus ("chapadas") corresponding to post-rift sedimentary covers preserved over the eroded rift structures. They result from topographic inversion of the post-rift basins (Peulvast et al. 2008). The inversion is stronger in the south (up to 450–500 m) than in the north (100 m or less between the Chapada do Apodi and the Apodi-Açu depression). It reflects the occurrence of differentiated uplift succeeding to the post-rift subsidence events.

Next to the Atlantic Ocean, the Chapada do Apodi presents a low, flat, and dry surface underlain by the Jandaira limestones, either shaped in karstic features (karren of Soledad: Fig. 3.4a) or covered by lateritized colluvium and, toward the coast, by the Cenozoic sediments (Tibau and Barreiras Formations) and the paleodunes that form the "tabuleiros." Gently sloping oceanward, down to 30–40 m along the coast, it is not entirely tabular. Around Mossoró, on both sides of the Rio Apodi valley, more elevated hills, domes or ridges (Serra do Mel, 45 × 30 km; Serra de Mossoró), corresponding to two short NE-trending anticlines, reach 125–270 m a.s.l. (Fig. 3.1). To the NE, the northeast tip of the Serra do Mel antiform is cut by the highest sea-cliffs of the northern coast of Brazil (100 m). Recently recognized as neotectonic features deforming the post-rift series (e.g., the tilted Jandaira limestone outcropping up to 200 m a.s.l., below a silicified remain of Barreiras sandstones), they represent local basin inversion along rift structures (Carnaubais-Portalegre shear zone) reactivated in a NW-SE to EW compressional stress field (Bezerra et al. 2011; Maia and Bezerra 2014b). Around these structures, the plateau roughly corresponds to a structural surface, except close to the southern rim, where erosion has beveled the limestone layers gently dipping toward the center of the basin (Fig. 3.2). The nature of the western and southern rims of the Chapada above the wide Jaguaribe, Apodi and Lajes orthoclinal depression—a cuesta shaped into the limestones and the underlying Açu sandstones (Fig. 3.4b)—as well as its dissection by a few consequent rivers also controlled by the NW–SE folds, reflect the topographic inversion of a broad and shallow, NE-trending synform, and respond to a slight uplift which probably accompanied its flexural tilting toward the margin and its more local compressional deformation (Fig. 3.5c, EF, GH, IJ, KL).

At the southwest end of the rift zone, the Chapada do Araripe stands as a high mesa, the surface of which approximately coincides with the uppermost, slightly W-tilted layers of the Exu sandstone (Peulvast and Bétard 2015). The nature of its dissected rims is that of an erosional scarp, dissected on both sides of the Chapada

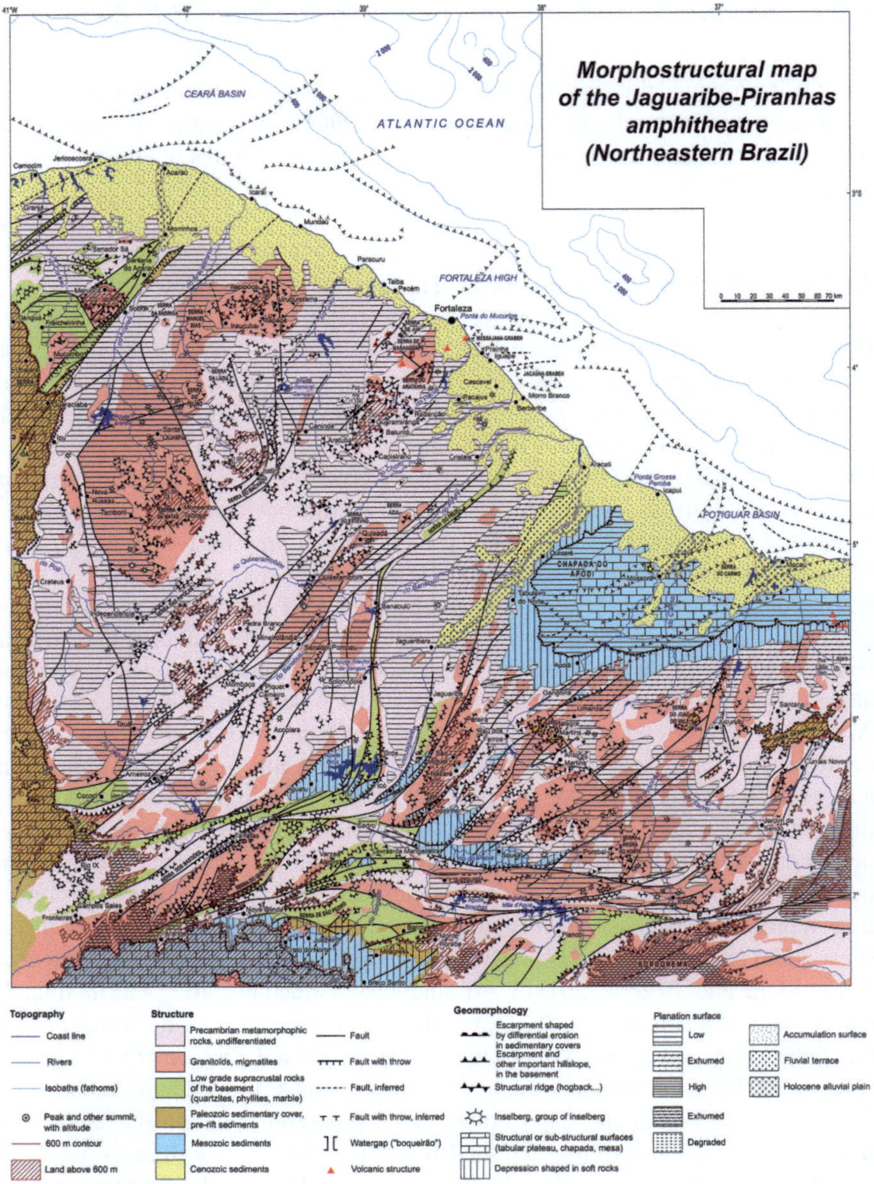

Fig. 3.2 Morphostructural map of the study area. Originally drawn on the Aeronautical chart ONC M-28 (scale 1:1,000,000) and elaborated from fieldwork and numerous sources including topographic and geological maps at various scales, Landsat images, radar mosaics, DEM (NOAA, SRTM 90), and air photographs

by wide box-canyons (Santana do Cariri, Jardim, Exu). Its formation reflects deep erosion by tributaries of the Jaguaribe and São Francisco rivers, and strong topographic inversion since the deposition of the post-rift cover. Although a rough

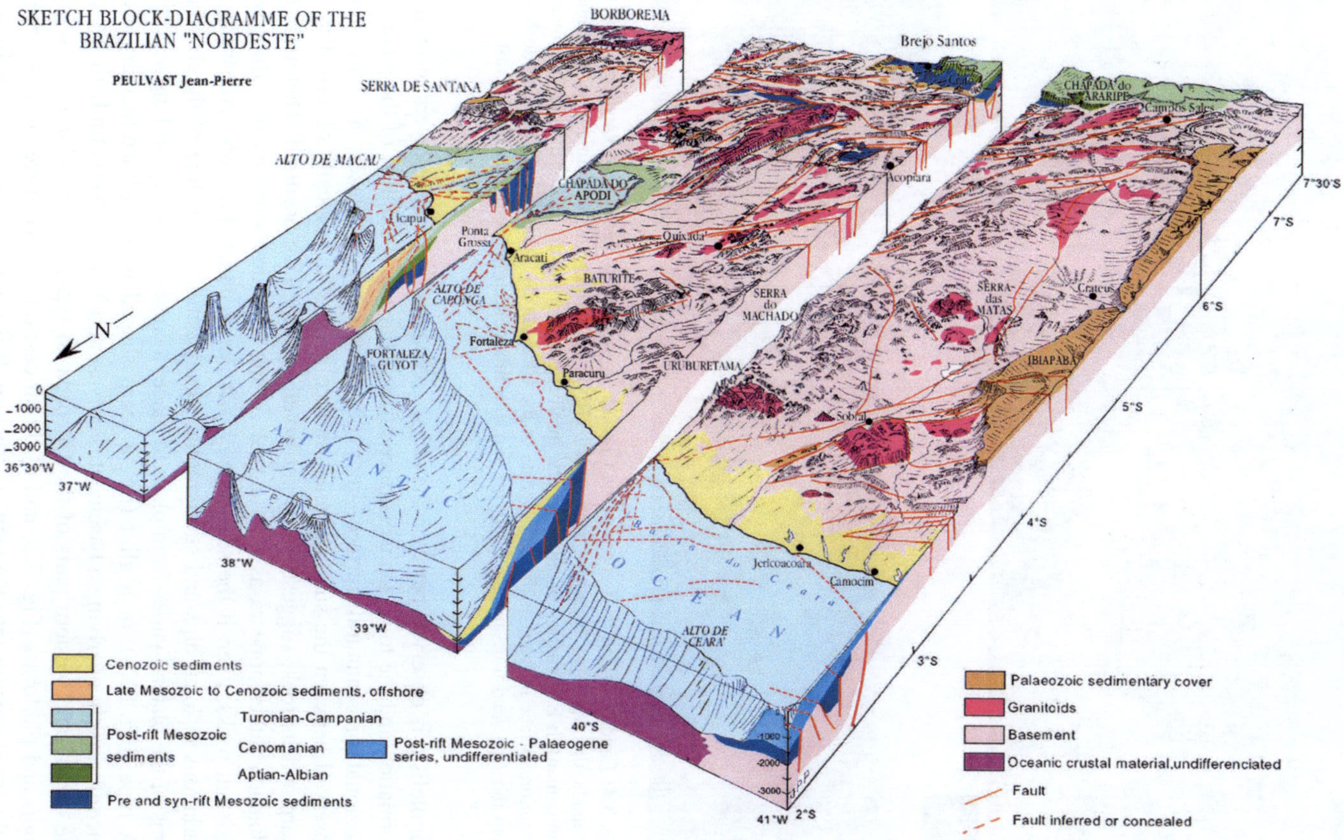

Fig. 3.3 Geological block-diagram of the equatorial margin of Northeast Brazil

Fig. 3.4 Landforms and landscapes of the Chapada do Apopdi. **a** Karren on a karstic plateau: the Chapada do Apodi at Soledad, near Apodi (RN). Flat-lying Jandaira limestone (Turonian-Campanian). **b** Exhumation of small granitic inselbergs or bornhardts and of the pre-Cenomanian planation surface in front of the west cuesta of the Chapada do Apodi, E of Limoeiro do Norte, from the Açu sandstone and the overlying Jandaira limestone. *Photographs* J.P. Peulvast

parallelism is observed between the deeply weathered sandstone layers of the Exu Formation and the topographic surface of the Chapada, the nature of the plateau—erosional or structural surface—may be questioned. In places, a slight angle observed between the sub-horizontal surface and the dips locally observed in the scarp and its rock, suggest that slight bevelling may have occurred. Therefore, the Chapada rather appears as a structural or sub-structural surface coinciding with resistant rocks, even if the sedimentary pile is not intact (Fig. 3.6). According to variations in local dips and stratigraphic superposition, the nature of the dissected rims of the plateau varies from place to place: a cuesta-like scarp on long segments (A type: Fig. 3.7a), or a glint (i.e., a scarp shaped into hard sedimentary rocks unconformably overlying a basement: Peulvast and Vanney 2001) on both sides of the central and western parts of the Chapada (B type), above exhumed basement highs and platforms (Fig. 3.7c and d). In all cases, this scarp owes its existence to differential erosion into the tabular structure. It is dissected on both sides by wide box-canyons or embayments continuing deep canyons (Fig. 3.7b). The initial

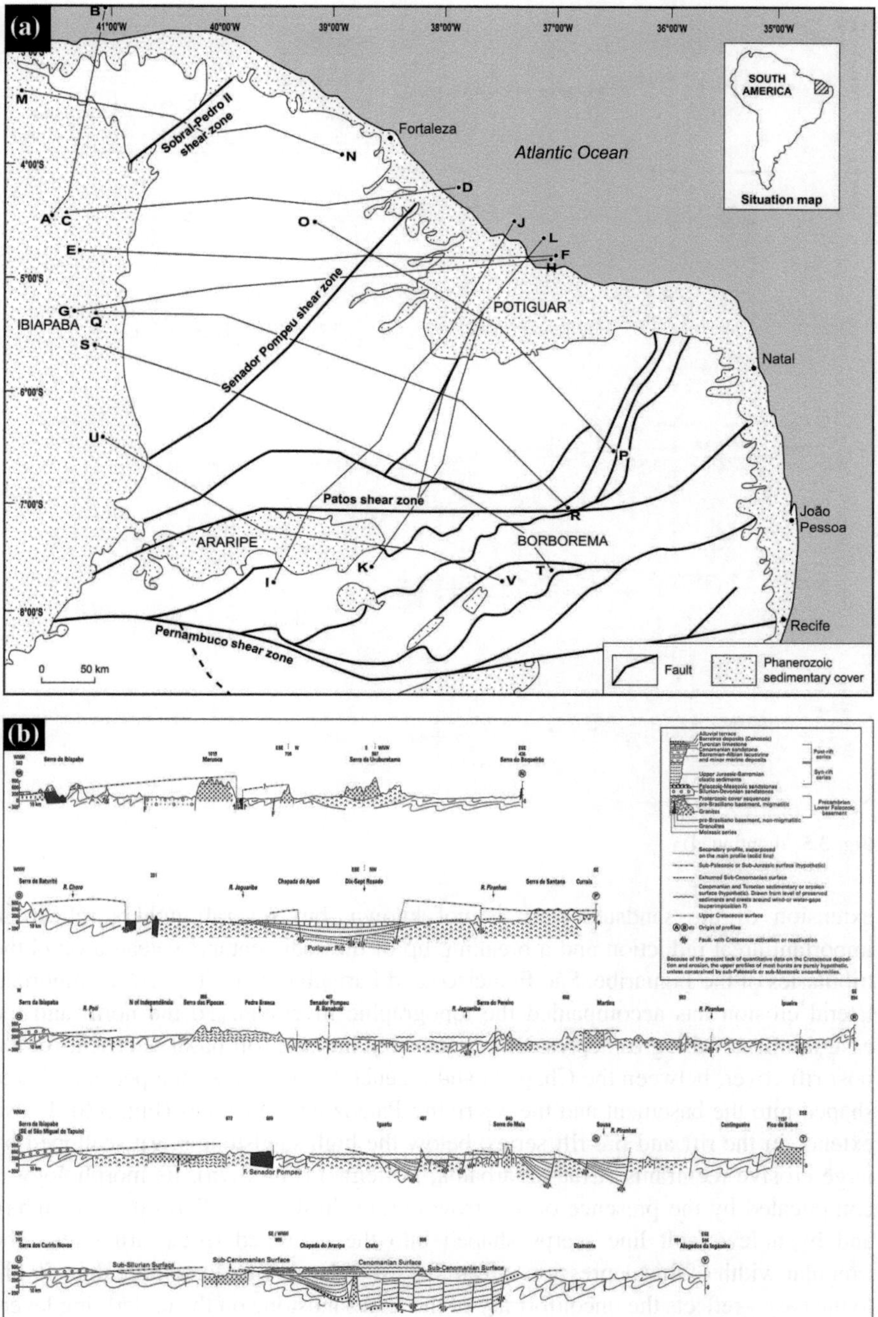

Fig. 3.5 Geological cross-sections through the study area. **a** Location map of the cross-sections. **b** Geological cross-sections through the Cariri-Potiguar rift zone. **c** Geological cross-sections normal to the margin

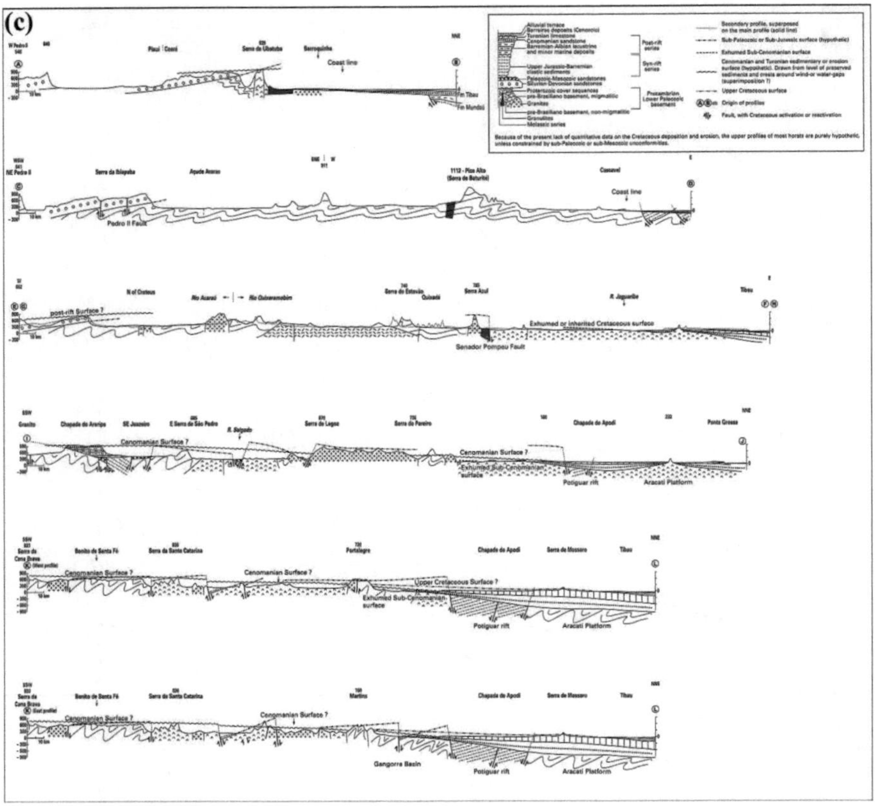

Fig. 3.5 (continued)

extension of the sandstone cap is not known, but several outliers reflect an important areal reduction and a breaking up of the sedimentary plateau around the tributaries of the Jaguaribe, São Francisco and Parnaiba rivers. Therefore, important lateral erosion has accompanied the topographic inversion. To the north and the east, the Crato-Milagres depression is excavated into the soft basal sediments of the post-rift cover, between the Chapada and a generally more elevated peripheral area shaped into the basement and the overlying Paleozoic sandstones (Fig. 3.6). It also extends in the rift and pre-rift series, below the high sandstone scarp scalloped by large erosive reentrants (Crato, Barbalha, Porteiras) (Fig. 3.7a). Its morphology is complicated by the presence of a narrow outlier, in the east (Serra da Mãozinha), and by a few fault line scarps shaped into the exhumed rifted structures. The irregular width of this depression—from 0 west of Nova Olinda to more than 50 km to the east—reflects the unconformity of the Exu sandstone on the underlying layers and the basement, and the maximum thickness and outcrop width of the rift and post-rift sediments in the Milagres–Brejo Santo area (see Peulvast and Bétard 2015, for more details about the morphostructural patterns of the Chapada do Araripe).

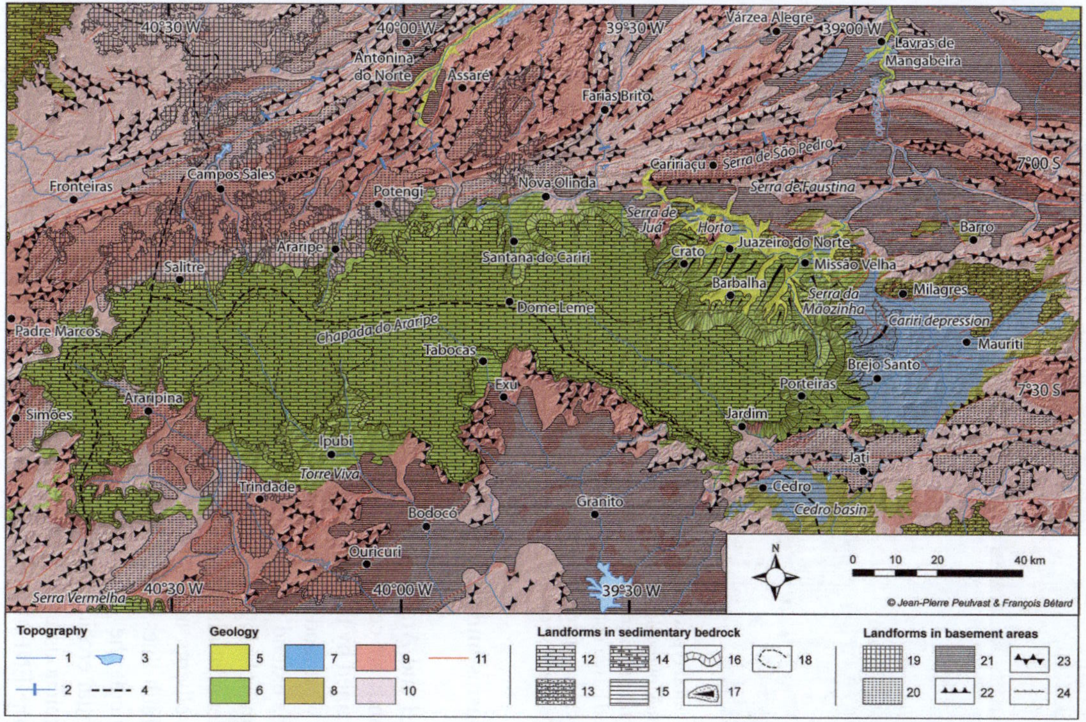

Fig. 3.6 Morphostructural map of the Chapada do Araripe and surroundings. *1* Main river; *2* knickzone; *3* lake ("açude"); *4* main drainage divide; *5* Quaternary alluvium; *6* post-rift series (Aptian to Cenomanian); *7* pre- and syn-rift series (Jurassic to Barremian); *8* Paleozoic sedimentary cover, pre-rift sediments; *9* granitoids, migmatites; *10* Precambrian metamorphic rocks, undifferentiated; *11* major fault; *12* structural or near-structural surface, sandstones of the Exu Fm (Cenomanian); *13* structural or near-structural surface, sandstones of the Mauriti Fm (Silurian?); *14* structural or near-structural surface, sandstones of the Serra Grande Fm (Paleozoic); *15* depression shaped by differential erosion into soft rocks of the rift and post-rift series; *16* escarpment shaped by differential erosion and sapping in tabular structures; *17* main remnants of pediments with colluvial cover, landslides and debris flows; *18* areas of shallow dissection in the structural surfaces; *19* high erosion surface (Step 1), with laterite; *20* elevated erosion surface, degraded or dissected; exhumed etch-surface at the periphery of vestigial duricrust mesas; *21* low erosion surface, often dissected by narrow and shallow valleys ("Sertaneja Surface"); *22* main escarpments; *23* main isolated ridges, crests and hogbacks; *24* edge of lateritic plateaus and mesas. From Peulvast and Bétard (2015)

Fig. 3.7 Morphostructural types along the rim of the Chapada do Araripe. **a** Cuesta-like scarp (A1 type) above Crato, fromMonte Alegre. Alternating landslide scars and sapping funnels. **b** Cuesta-like scarp with sandstone outliers overlooking the sedimentary depression of Ipubi (A2 type). **c** Glint over exhumed and lateritized basement surface near Salitre (B1 type); complicated pattern of stepped rock faces to the west (*right side*): effects of mass movements (large slumps); **d** Glint and composite scarp west of Exu (sandstone crag over granite bench; B2 type). *Photographs* J.P. Peulvast

3.1.1.2 Horsts, Basins and Escarpments

Between the Apodi and Araripe plateaus, two major depressions belonging to the inner ring of lowlands are centered on Neocomian half-grabens deprived of post-rift cover. They are drained by the Jaguaribe (Iguatu-Icó basin) and Piranhas rivers (Rio do Peixe basin), west and south of the Pereiro-Antonio Martins horst system. Smaller depressions extend within this massif around the grabens of Pau dos Ferros and Nazaré. The morphology of intervening horsts and escarpments displays evidence of strong denudation (Fig. 3.5c, IJ, KL). Most outlines of the massifs, inselgebirge and inselbergs are sinuous and/or discontinuous but can be related to nearby or remote fault lines, or to lithological contacts, or to both types of controls, often making difficult the distinction between tectonic scarps and those purely shaped by differential erosion (Peulvast et al. 2006; Gurgel et al. 2013).

The most conspicuous escarpment is the western edge of the Pereiro granitic massif, related to the ENE-trending Jaguaribe fault over more than 100 km. Except for a wide embayment in the south, its southern half is a straight or *en échelon* granitic wall, only dissected by a few hanging valleys, which overlooks the Icó half graben, 2–5 km east of its master fault line. It appears as a fault scarp which has retreated to the intrusive contact parallel to the fault zone, in relation with the development of rugged pediments (1–5 km) and longitudinal corridors in the footwall (Fig. 3.5b, QR). Although it is interpreted as a rejuvenated fault scarp by Gurgel et al. (2013; see discussion in Sect. 5.2.3), the presence of high residual basement hills or crests on the hanging wall suggests that exhumation and denudation by downwearing of this compartment have also contributed to scarp formation. Such a situation probably also occurred to the north, where the escarpment and the tip of the Pereiro massif are subdivided in several parallel gneissic and granitic ridges or elongated domes between branches of the shear zone, with sharp knicks or short concavities at the contact with the Sertaneja surface. Tectonic controls are more complex than in the south, between the Jaguaribe and Apodi depressions. Digitated outlines are those of residual horsts or half horsts corresponding to resistant granitic intrusions, 10–40 km from the nearest master faults, respectively the Orós-Jaguaribe fault system to the west and the buried Apodi transfer fault to the NE (Fig. 3.2).

Structural controls on other major scarps are also ambiguous. The Rio do Peixe and Pau dos Ferros basins are surrounded by remote (5–15 km) sinuous escarpments or alignments of dome-shaped inselbergs on hanging walls and foot walls as well (Fig. 3.3). The most conspicuous fault-related landforms are fault line valleys and basins hollowed out by differential erosion into the intervening horsts, in particular along the Portalegre and Nazaré fault zones. In some of the largest basins (Icó, Antenor Navarro-Sousa), low fault line scarps (up to 10–30 m) are shaped along the south or southeast sides of the main border faults, owing to easy excavation of the soft basin sediments (Fig. 3.8a). Related to recent entrenchment of the rivers into the Sertaneja surface, they separate the slightly depressed alluvial floors from this plain. Here, the Sertaneja surface is surrounded by sinuous escarpments and scattered inselbergs representing remnants of the uplifted foot walls (Cajazeiras and Icó areas). Contacts between basin floors and highlands are more progressive on the hanging walls, on the north or northwest sides of the Rio do Peixe and Iguatu basins.

Because of the lack of post-rift cover in these basins, the amount of erosion at the expense of the rift sediments is unknown, so that the nature of the surrounding slopes—residual or exhumed fault and flexure scarps—is difficult to ascertain. The small volumes of sediments preserved in some of them, in the middle of wide depressions (Pau dos Ferros) or their presence in deep fluvial valleys (Tarrafas) suggest that they only represent remnants of once thicker and larger deposits. Although the fine preservation of dinosaur prints and dessiccation craks in flat-lying mudstones of the Sousa Formation in the Rio do Peixe basin, presently exhumed by river erosion near Sousa, does not suggest strong compaction by a former thick overburden (Fig. 3.8b), it appears that the final syn-rift filling of the basin was probably thicker (Sénant and Popoff 1991). However, the small size and depth of these intermediate basins, their location between high standing and wide horsts, and the lack of any remnant of such a cover, only found where capped by resistant layers in nearby basins, does not favor the hypothesis of wide post-rift subsidence and deposition. North of Portalegre, the wide onlap of post-rift sediments over the southern shoulder of the Potiguar rift (Figs. 3.3 and 3.5c, IJ, KL) shows that significant syn-rift erosion had already occurred on the uplifted basement and probably in the rift basins outside the last depocenters.

Whatever be the origin of the Sertaneja surface in these basins—syn-rift or later erosion, it appears as the lowest significant step of the local topography, below the

Fig. 3.8 The Rio do Peixe basins. **a** The south rim of the Rio do Peixe (Antenor Navarro-Sousa) basin between Sousa and Cajazeiras (PA). Low fault line scarp (10–30 m) shaped along the main border fault of the Neocomian half-graben, owing to easy excavation of the soft basin sediments in relation with recent entrenchment of the river into the planation surface which had previously cut both the basement and the rift fill (Sertaneja surface). **b** Dinosaur prints in the Rio do Peixe basin near Sousa (PA). Mudstones of fluvial-lacustrine origin (Sousa Formation, Neocomian). Fine exhumation of weakly tilted strata having preserved fresh looking desiccation features. *Photographs* J.P. Peulvast

flat or rolling surfaces that form the top of the widest massifs, between 650 and 850 m a.s.l. Rolling etch surfaces (Thomas 1989a, b) strewn with rocky knobs, bornhardts and tors, between irregular etch basins blanketed with thick kaolinitic saprolites and soils, form wide areas of the granitic Pereiro plateau and of the dissected Portalegre-Antonio-Martins massif (Bétard 2007). They bear scattered remnants of lateritized saprolites and fluvial sandstones (Serra do Martins Formation), forming flat buttes (Pereiro: Fig. 3.9) or mesas (Portalegre, Martins, João do Vale).

3.1.1.3 The Araripe Block and Basin

South of the Upper Jaguaribe and Rio do Peixe Rivers, the wide band of basement highlands that extends between the Iguatu-Rio do Peixe basins and the Araripe mesa also bears narrow remnants of high planation surfaces (Serras dos Bastiões, da Quincunca). They are well preserved to the west, around Campos Sales and Pio IX, owing to protection by an extensive lateritic cap (Fig. 3.6). Their altitudes—600 to 800 m—are those of the basal unconformities of nearby flat-lying sedimentary

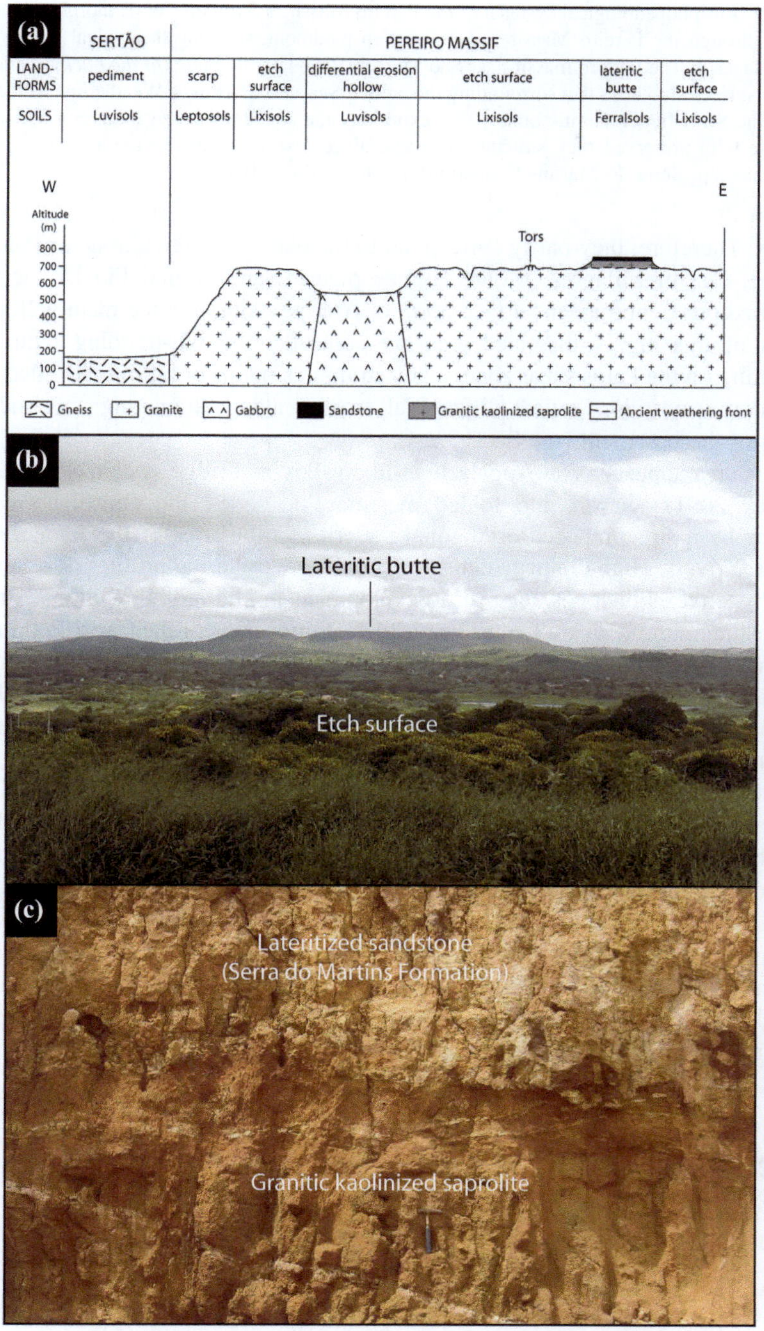
(a)
SERTÃO
PEREIRO MASSIF
LAND-FORMS
pediment
scarp
etch surface
differential erosion hollow
etch surface
lateritic butte
etch surface
SOILS
Luvisols
Leptosols
Lixisols
Luvisols
Lixisols
Ferralsols
Lixisols
W
E
Altitude (m)
Tors
Gneiss
Granite
Gabbro
Sandstone
Granitic kaolinized saprolite
Ancient weathering front
(b)
Lateritic butte
Etch surface
(c)
Lateritized sandstone (Serra do Martins Formation)
Granitic kaolinized saprolite

◀ **Fig. 3.9** Morphopedological systems of the Pereiro Massif. **a** Synthetic W–E morphopedological profile through the Pereiro Massif and its western piedmont, showing the typical soil-landform assemblages. **b** The Pereiro massif near São Miguel (RN): lateritic butte (*at the back*) overlooking an etch surface similar to that surrounding the eroded Serra do Martins. **c** Weathering profile at the top of the São Miguel lateritic butte (RN), exhibiting the contact between the granitic kaolinized saprolite with preserved rock structure (coarse-grained texture, quartz veins) and the lateritized sandstone cap (Serra do Martins Formation). *Photographs* F. Bétard

covers. Therefore, they partly correspond to the exhumed pre-Jurassic or Paleozoic surface, east of Milagres (Monte Horebe plateau) and around Pio IX, or to the exhumed pre-Cenomanian surface north of Araripe. Although the main valleys and ridges of this highly dissected area are controlled by NE-trending fault zones branching on the Patos shear zone, many straight, sinuous or hairpin bevelled crests and narrow corridors reflect differential erosion along alternating soft and hard layers of the basement (see Sect. 3.2.4).

This area appears as an Appalachian-like valley-and-ridge system (Peulvast and Vanney 2001), shaped into folded metamorphic series and Brasiliano granitic plutons by a superimposed orthogonal pattern of tributaries of the Jaguaribe and Parnaiba rivers. High drainage densities and deep weathering profiles developed on mainly convex hillslopes suggest a development in "humid" conditions, strongly contrasting with the overall concave profiles of drier parts of the Cariri-Potiguar rift zone and of the Borborema region. Similar drainage patterns are observed on the tributaries of the Piranhas River, which flow northwards from the Monte Horebe plateau down to the Rio do Peixe basins, through EW metamorphic and granitic residual ridges rising on the Sertaneja surface.

The presence of narrow grabens still partly filled with pre-rift sandstones (Lavras da Mangabeira, Barro) suggests that selective erosion related to valley entrenchment into the basement is not the only cause of relief differentiation. Tectonic disruption related to Early Cretaceous rifting also played a role, although this area, uplifted along the Patos shear zone acting as a transfer fault between the Rio do Peixe and Araripe basins, remained rather massive. Later erosion related to shaping of the Sertaneja surface and hollowing out of the Crato-Milagres inner lowland allowed partial exhumation of uplifted or tilted unconformity surfaces preserved on the hanging walls of some basins, below high surfaces of various origins (Fig. 3.5b, UV).

3.1.2 The Northwest Shoulder of the Cariri-Potiguar Rift Zone

3.1.2.1 A Tectonic Block?

The parts of the highland and depression rings that extend west of the Jaguaribe depression correspond to the north-west shoulder of the Cariri-Potiguar rift zone. The absence of any identified faulted basin—except for narrow Paleozoic sandstone bodies downfaulted along the Transbrasiliano shear zone near Santana do Acarau—suggests that the Central Ceará highlands belong to a wide NS or NNE–SSW block, the Santa Quiteria-Tamboril-Baturité domain, uplifted between the Cariri-Potiguar rift zone and the Parnaiba basin (Fig. 3.1). To the south, this block joins the inverted Araripe basin without any indication of post-rift reactivation of fault structures. All the fault lines of the Central Ceará and Orós-Jaguaribe basement units, and of the Patos shear zone are sealed by the Cenomanian and Silurian-Devonian sandstone covers of the Araripe and Parnaiba basins, or truncated by the exhumed planation surface that prolongs the basal unconformities of both sedimentary units around Campos Sales (Fig. 3.5b, UV). North of this area, the block is widely excavated below the rim of the Parnaiba basin.

Morphotectonic disrupting of this block only appears north of the Patos shear zone, in two limited areas. South of Acopiara, the southeast edge of the uplifted block corresponds to the hanging wall of the Iguatu basin. On the northwest side, along the Transbrasiliano shear zone, deformations of Devonian sandstones preserved in grabens and half-grabens (Destro et al. 1994; Galvão and Da Silva 2002) suggest that the edge of this huge block corresponds to the wall of this narrow downfaulted zone. Vertical movements along this structure were more important in the north, near the ocean, than in the south, where vertical offsets decrease in the Parnaiba basin. On both sides of this block, divergent dips of sedimentary covers (Potiguar and Parnaiba basins) suggest a low dome-shape for this huge structure corresponding to the Central Ceará domain of the basement, probably with a wide flat top (Fig. 3.5b, QR, ST and c, CD, EF, GH). Westward tilting only increases and becomes regular west of the Serra da Ibiapaba (Parnaiba basin).

3.1.2.2 The Topographic Edge

North of the Iguatu basin, no clear relationships appear between the eastern slope of the Central Ceará highlands and the main tectonic structures, the remote fault zones of the Orós-Jaguaribe unit, presently bevelled by the Sertaneja surface in the Jaguaribe plain. The remarkable width and regularity of this surface—with the exception of long series of quartzitic hogbacks (Banabuiú-Orós and Serra do Felix ridges)—and its eastward continuation in the exhumed basal unconformity of the post-rift cover of the Potiguar basin (Fig. 3.5b, OP) suggest that it was shaped into a wide flexural slope partly corresponding to the hanging wall of the Potiguar and Iguatu basins. Cretaceous or later faulting may have occurred along former Brasiliana shear zones (e.g. the Senador Pompeu fault zone; Figs. 3.1 and 3.3), but no specific landform such as a fault scarp could be identified.

The existence of a wide flexure is suggested by the morphology of the east edge of the Central Ceará highlands. The 300 km long, approximately NS *en échelon* series of promontories and embayments that forms a very irregular and discontinuous escarpment (here called Acopiara-Baturité escarpment) above the Jaguaribe plain displays various types of structural controls. South of Mombaça, the Sertaneja surface bevels the hanging wall of the Iguatu-Orós fault zone over a width of

10–20 km, south of the low migmatitic Acopiara promontory (Fig. 3.5b, ST). A long NNE-trending row of granitic inselbergs and inselgebirge prolongs the Acopiara massif to the north, forming the west edge of this block between the Orós and Senador Pompeu fault zones. This residual ridge partly coincides with a syn-kinematic granitic intrusion, north of Piquet Carneiro, whereas another elongated granitic intrusion located to the east along the Orós-Banabuiu fault zone was bevelled by the Sertaneja surface and then dissected (Fig. 3.2).

North of Mombaça, a possible influence of Cretaceous movements along the Senador Pompeu fault zone may be suspected, in spite of increasing distances between this shear zone and the Ceará highlands. Here, the northwest shoulder of the Cariri-Potiguar rift zone appears strongly eroded. Wide (more than 25 km) embayments, deeply eroded into the metamorphic rocks that form these highlands, are connected to the Jaguaribe plain, west of Quixeramobim and south of the Baturité massif (Fig. 3.3). In these areas, the east edge of the highlands is located 30–60 km away from the Senador Pompeu shear zone, where no clear evidence of post-Paleozoic movements is observed. However, the presence of numerous granitic inselbergs and inselgebirge (Quixadá, Serra Azul) within the shear zone and on its northwest side—and the lack of comparable landforms on the opposite side, the mainly gneissic Orós-Jaguaribe domain—suggest that this escarpment might be a residual fault scarp (Peulvast and Vanney 2001) shaped by irregular dismantling or retreat above the Sertaneja planation surface, which developed less easily here than into the downfaulted block. It might also result as well from differential erosion, without differential uplift, between the syn-orogenic granites and homogeneous gneiss of the Orós-Jaguaribe domain, to the southeast, and more resistant metamorphic rocks of the Ceará and Independência Groups to the west.

Structural control on this escarpement is better identified between the Senador Pompeu and Quixadá areas, where the Quixadá-Quixeramobim monzonitic complex, intruded between the parallel Senador Pompeu and Quixeramobim shear zones, is partly truncated by the Sertaneja surface (Fig. 3.2). The metamorphic basement (micaschists, quartzites, gneisses, amphibolites) forms the hills and ridges that continue the Central Ceará escarpment to the northeast, on the northwest side of the Quixeramobim shear zone (Serra do Estevão; Fig. 3.5c, EF, GH). In the calc-alkaline granitic complex, i.e., on the northwest side of the Senador Pompeu shear zone, groups of bornhardts, ridges, and spectacular inselbergs, up to 350 m high, are preserved in leucocratic facies (Fig. 3.10a), whereas in the ovoid Quixadá intrusion (260 km^2), the "mar de inselbergs" more or less underlines the circular internal foliation of the batholith and the fracture patterns (Fig. 3.10b; see Sect. 3.2.4). On the contrary, most intermediate and basic facies are eroded (diorites and quartz-diorites, granodiorites). Together with the occurrence of the Serra Azul granitic inselgebirge along the same shear zone, 40 km NE of Quixadá, this is the only case where a lithological control is clearly identified along this escarpment.

From Quixadá to the coastal area, structural controls on the escarpment outlines are less clear, especially on the east side of the Baturité massif, which turns to NNE, reaching the southern outskirts of Fortaleza. Here, more than 50 km to the west of the Senador Pompeu shear zone, no continuous fault line is mapped or even visible on radar and 3D images (Projeto RadamBrasil 1981a, b), neither between this shear zone and the sinuous escarpment that overlooks the Sertaneja pediplain, nor along the foot of the escarpment. The main massif, in the south, is mainly shaped into supracrustal rocks (the Independência and Canindé Formations of the Ceará Group: CPRM 2003; Bétard 2007). From Capistrano to Redenção, several features suggest the erosive origin of its eastern border: the irregular and sinuous outlines of this 500–600 m high hillslope, including short and steep valleys opening on five large embayments, the presence of a long NE-trending promontory, the local lithological controls on the shapes and position of the hillslopes (domes shaped into massive gneissic layers, marble and dolomite crests and pinnacles, quartzitic hogbacks), and

Fig. 3.10 Inselbergs and structural controls. **a** Quixadá (central Ceará): Serra da Estrela, Serra do Mel, as seen toward SW from the Serra do Urucu. Groups of bornhardts, ridges, and spectacular inselbergs, up to 500 m high, shaped into the leucocratic facies of the Quixadá-Quixeramobim granitic complex. They rise between structurally controlled corridors (intermediate and basic facies: diorites and quartz-diorites, granodiorites; fractures branching on the parallel Senador Pompeu and Quixeramobim shear zones) connected to the Sertaneja surface. **b** "Mar de inselbergs" shaped into the ovoid Quixadá monzonitic intrusion, as seen to the north from the Serra do Urucu. The granitic inselbergs protrude in front of the east edge of the Central Ceará highlands above the Sertaneja surface. In the distance: residual reliefs preceding the eastern escarpment of the Baturité massif. *Photographs* J.P. Peulvast

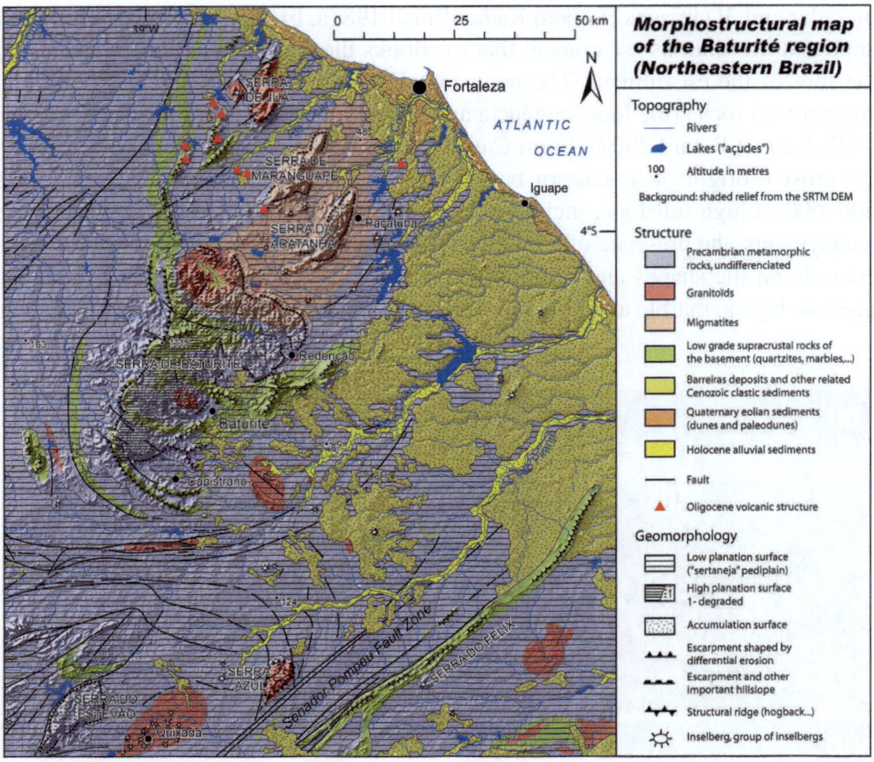

Fig. 3.11 Morphostructural map of the Baturité massif and surroundings

the presence of several high inselbergs in front of it, north of the Serra Azul (the aplitic Pedra Aguda; Fig. 3.11).

Although large parts of the Baturité massif are underlain by the same metamorphic rocks that form the surrounding pediplains (i.e., the Canindé Formation), most peaks, culminating ridges, and outstanding escarpments are shaped into the folded quartzite layers of the Independência Formation, which form the real bone structure of the massif and multiply the thresholds of hard rocks preserving the shallow valleys of the plateau from regressive erosion (Fig. 3.11). Therefore, the outlines may be considered as those of a residual massif, although the absence of residual reliefs on the opposite side of the Senador Pompeu shear zone suggest that a fault scarp could have retreated from this fault during or after the Cretaceous rifting, into the uplifted compartment. To the northeast, both compartments of the same fault are bevelled by the same surface that disappears below the Cenozoic sediments of the coastal area and truncates the buried rift structures of the margin, excluding any significant movements in recent times (Late Cenozoic). In fact, the outer tectonic limits of the eroded compartment may be those of the Aracati platform (Fig. 3.5b, OP, and c, CD).

A more local tectonic control may be suspected north of Redenção, on the eastern side of the Serra da Aratanha, one of the northern outliers of the Baturité massif, shaped into the thick migmatitic and granitic nappe that is also partly bevelled by the Sertaneja surface to the southwest of Fortaleza. Above Pacatuba, its straight outlines and its steep rectilinear profile look like those of a high fault scarp. But here again, the escarpment is not entirely located on the lithological contact with the underlying gneisses, and lies 3 km to the west of an N20 trending fault line that controls the shallow upper Rio Cocó valley (Pacatuba-Açude Gavião), south of Fortaleza (Fig. 3.11).

3.1.2.3 Western Highlands and Depressions

To the west of the Acopiara-Baturité escarpment, the highlands that form a 200–250 km wide NS zone in central Ceará are more and more dislocated northwards, finally forming isolated massifs aligned in the western part of the marginal escarpment. Several morphological types are identified here. Wide arrays of elongated, often arcuate (Serra de Santa Luzia, Serra do Machado) ridges of quartzite and other metamorphic rocks of the Ceará and Indepêndencia Groups, trending NS to NE–SW, form the bulk of these highlands, together with dissected plateaus and jagged systems of peaks and crests rising hundreds of meters above their pedestal of irregular high plains (Fig. 3.3). Structural ridges and valleys, narrow tabular surfaces, high gneiss or quartzite ledges, and flat-floored depressions are controlled by tight networks of folds and orthogonal or curved fault lines and fractures. Some of the highest hogbacks and escarpments correspond to Brasiliano thrust zones in the gneiss complex and in the overlying basement nappes (west-facing escarpments of the orthogneissic Pedra Branca plateau and Serra das Pipocas). Some of them overlook the surrounding Sertaneja erosion surface cut into metasediments (Indepêndencia area), whereas in other places this plain forms irregular embayments in the dissected pedestal (east flank of the Central Ceará highlands).

The Baturité massif differs from these highlands by its massive morphology of dissected plateau (Bétard 2007, 2012). The central, summit part of the massif is characterized by constant elevations around 800–900 m a.s.l., only interrupted by elongated crests and quartzitic ridges including the highest points above 1000 m a.s.l. (Pico Alto, 1115 m). Corresponding to the remnant of an ancient erosional level, probably pre-rift in age, this summit surface is characterized by multiconvex topography (*meia-laranjas* or convex hills) shaped into thick saprolites derived from the weathering of gneisses and granites under conditions of humid tropical climate (Fig. 3.12a). The peripheral area of the massif is typified by highly dissected landscapes (200–800 m a.s.l.), bearing strong marks of active dissection and regressive erosion from the dense river network flowing to the piedmont (Fig. 3.11). Orthogonal patterns of steep and narrow canyons, hanging valleys and depressions of differential erosion excavated at various levels dissect all the borders of this gneissic plateau in sharp crests (west flank), rounded domes (Baturité), homoclinal ridges and hogbacks of quartzite (south, west). High peaks and domes rise over the

Fig. 3.12 Contrasted landscapes of the Baturité-Irauçuba compartment. **a** Deep weathering profile and associated Acrisol on a forested convex hill, at the top of the humid Baturité massif. **b** Granitic domes and needles in the semi-arid landscape of the Itapagé valley (Irauçuba massif). *Photographs* F. Bétard and J.P. Peulvast

lower and more dissected plateau that forms the northern edge of the massif, carved into gneisses, granites and migmatites, above a wide corridor excavated in the same migmatites, between the Baturité and Aratanha mountains.

To the west, similar landforms characterize the Uruburetama massif, in spite of deeper dissection. Jagged granitic and migmatitic domes and peaks (Fig. 3.12b) rise between narrow valleys and pediments connected to inner depressions—i.e., the long EW corridor hollowed out along the south edge of the central granitic bastion, and to surrounding plains. Unlike the Baturité plateau, this massif is surrounded by numerous ridges and inselbergs, especially in the southern and western metamorphic envelopes of the massif, where semi-arid conditions and locally active regolith stripping reinforce the rigidity of the profiles of pediments and knicks.

In both cases, no systematic tectonic control (fault line, fracture) is identified in the outer limits of the mountains. The role of differential erosion is best ascertained in the Uruburetama massif, whose outlines are approximately those of a thick crystalline klippe preserved in a synform over the gneissic units of the Ceará Group (Fig. 3.5b, MN). On the contrary, the similar unit which is preserved south of Fortaleza is widely bevelled by peripheral and inner depressions and pediments. It only bears narrow but high and jagged residual ridges in the northern prolongation of the Baturité massif (Serra da Aratanha, Serra de Maranguape, Serra de Jua, Serra da Conceição, up to 920 m a.s.l.).

More abruptly delimited massifs form some of the other summits of the central Ceará highlands. The highest of them towers on the western edge of these mountains, between Monsenhor Tabosa and Tamboril. The outer wall slopes and dome-shaped flanks of this circular granitic bastion named Serra das Matas (1124 m) coincide with the outer limits of this late-Brasiliano granitic intrusion (Fig. 3.3). Nevertheless, it is highly dissected by an orthogonal pattern of fault line U- or V-shaped valleys and surrounded by lower uplands (the Monsenhor Tabosa plateau: 700 m), ridges and inselbergs shaped into metamorphic rocks and other intrusions, south of Santa Quiteria.

On the contrary, the quadrangular Serra da Meruoca (23 × 28 km, 1020 m) directly overlooks the Sertaneja pediplain west of Sobral. Strongly dissected by an orthogonal pattern of shallow hanging valleys, this high granitic plateau turns straight and abrupt escarpments respectively toward the Jaibaras graben and its clastic, volcanoclastic and volcanic infilling (SE), the scattered hogbacks and inselbergs of the metamorphic Senador Sá uplands (NE), and the embayments of the Sertaneja surface that are excavated into the molassic rocks of the Ubajara basin, between the massif and the piedmont hills of the Serra da Ibiapaba (NW, SW). The bases of all these escarpments coincide with orthogonal fault lines along which the granite was intruded (Fig. 3.2).

The highest escarpment (SE) coincides with the Café-Ipueiras fault of the Transbrasiliano shear zone, along which the conglomerates of the Aprazivel Formation of the Jaibaras graben are partially verticalized (Teixeira et al. 2004). Its triangular facets look like those of a fresh fault scarp (see discussion in Chap. 5, on the possible influence of neotectonics). The other escarpments face depressions and high quartzitic residual ridges whose preservation suggests that differential erosion occurred between the granite and the surrounding arkoses, graywackes, slates, and limestones of the Ubajara basin (Fig. 3.3). Therefore, they may be defined as fault line scarps. Moreover, the deep erosion of the nearby Mucambo granitic intrusion, reduced to a few inselbergs and a narrow arcuate hill (600 m) in front of the Ibiapaba escarpement (Fig. 3.13a), shows the importance of the denudation that has occurred on the NW shoulder of the Transbrasiliano shear zone. In addition to 300–400 m of the unconformable Paleozoic sandstones of the Serra Grande Formation, at least 400–500 m of basement rocks were removed from this block. The difference in resistance of the twin intrusions to erosion may be less important than suggested by the difference in altitude, because of the dip of the unconformity surface that once truncated them (Fig. 3.13b), but differences in composition (syeno-granites and alcaline feldspar granites in the Meruoca massif, quartz syenites, and quartz monzonites in the Mucambo pluton) or differences in outcrop level may also be involved. Strong cataclasis and hydrothermal alteration were also found along the contact between the Meruoca granite and the Jaibaras rocks (Teixeira et al. 2004; Santos et al. 2013). Like the Pereiro and Baturité massifs, the Meruoca intrusion is bevelled by a rolling surface dissected by an orthogonal pattern of hanging valleys, between convex hills with thick weathering profiles. Geological cross sections suggest that it may derive from the exhumed pre-Devonian surface (Fig. 3.5b, MN).

The large Canindé depression, which extends 110 km inland from the coastal area along a NS or NNE–SSW direction, between the Baturité and Uruburetama massifs, is mainly hollowed into the metamorphic cover sequences of the Ceará

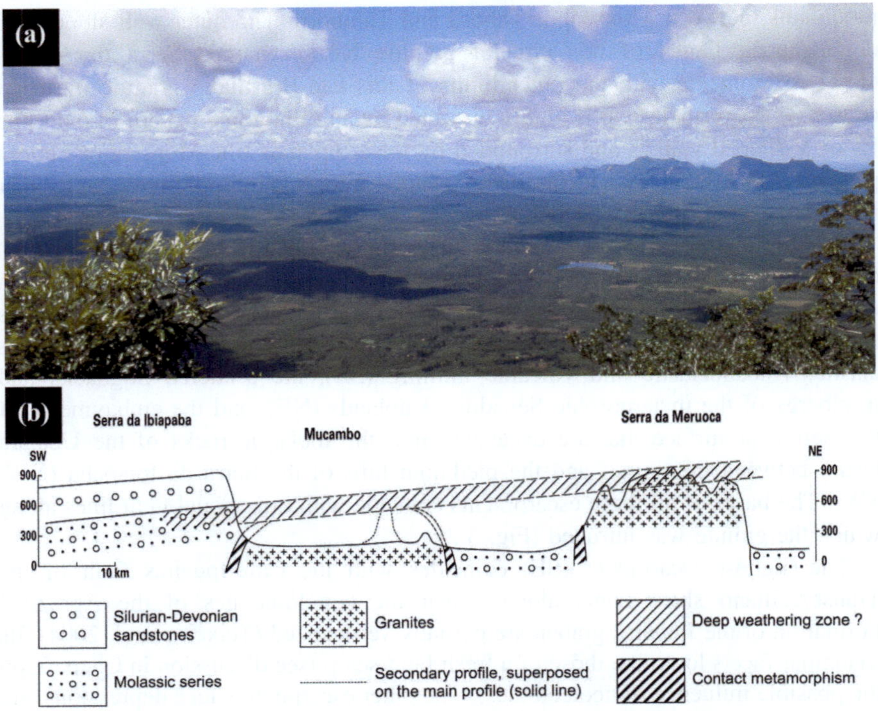

Fig. 3.13 Late-Brasiliano granites and differential erosion. **a** The Meruoca et Mucambo sister intrusions, as seen to the east from the Ibiapaba glint at Ubajara (NW Ceará). Whereas the Meruoca intrusion (*at the back*) bears a high rolling surface (900–1000 m) surrounded by escarpments coinciding with the orthogonal fault lines along which the granite was intruded, the Mucambo intrusion is strongly hollowed out and reduced to a few inselbergs and a narrow arcuate hill (600 m) at the foot of the Ibiapaba escarpement. The difference in resistance of the two granitic intrusions to erosion may be less important than suggested by the difference in altitude, because of the dip of the unconformity surface that once truncated them (**b**). *Photograph* J.P. Peulvast. **b** Geological cross-section through the Ibiapaba, Mucambo and Meruoca mountains

Group (Canindé unit). Clear limits are only visible on its east side, with the 50 km long quartzitic escarpment of the Baturité massif (Fig. 3.11). Only a few hills, inselbergs and dissected promontories of intermediate altitude protrude between short and steep gullies above the pediments developed in front of this dissected escarpment. Its sinuous outlines are only locally controlled by a few NE trending fault lines. North of Itapebuçu, similar control is observed along the westernmost units that prolong the Baturité massif to the north, a system of NS alignments of scattered quartzitic ridges (Independência Unit) and Oligocene volcanic necks surrounded by dissected pediments and lowering to the coast (Fig. 3.11).

Big inselbergs are preserved above the dissected floor of the depression at its southern end, around Canindé. Arcuate quartzitic ridges of the Independência Unit surround the digitated embayment that indents its west rim between General Sampaio and the upper valley of the Curu River. The crests and inselbergs shaped into the metamorphic envelop of the Uruburetama massif, and the easternmost ridges of the Serra do Machado (940 m a.s.l.), formed by quartzites and other metasedimentary rocks of the Independência Unit, also show lithological controls. They overlook systems of low ridges and hills where remains of high fluvial terraces (around the Curu river), and dissected pediments progressively merging with the floor of the depression near Canindé reflect the last planation and incision stages (Peulvast and Claudino Sales 2004).

3.1.2.4 The Outer Depression and the Western Glint

The only part of the outer depression that may be controlled by Paleozoic or later faulting is the north part of the Santa Quiteria plain, bounded to NW by the Transbrasiliano shear zone and the Santana do Acaraú graben (Fig. 3.2). These structures control the middle reach of the Acaraú river, a river whose strongly asymmetric catchment underlines the asymmetric profile of the depression, with a sharp contrast between its straight, but discontinuous, west wall and its remote, sinuous, and strongly dissected east rim (Fig. 3.3).

The gneissic floor of the depression gently slopes to NW, from the Central Ceará Highlands to the upper Acaraú valley, slightly entrenched upstream of Sobral. High granitic or migmatitic inselgebirge (729–947 m a.s.l.) and scattered or aligned inselbergs shaped along NW–SE directions in various basement rocks and quartz dykes tower above this surface around Santa Quiteria (Fig. 3.3). The depression narrows southwards, from 80 to 40 km (Nova Russas), between the Monsenhor Tabosa mountains and the Serra da Ibiapaba, but preserves its fundamental EW asymmetry. Here, faint granitic hills form a low divide (360–600 m) between the Santa Quiteria and Crateus depressions.

Like the Crateus depression and the more symmetric depression of the Upper Jaguaribe river, the whole of the very long (300 km from north to south) and wide (50–80 km) western corridor is hollowed into the gneissic and migmatitic basement (Ceará Group), below the uplifted post-Brasiliana unconformity surface (Fig. 3.5c, EF, GH). The hollowing depth decreases from north to south (500 m or more to 100–200 m), although a second maximum is reached around the Poti river, the upper reaches and tributaries of which are entrenched into the floor of the Crateus depression. This depth is generally maximal at the foot of the Ibiapaba escarpment or glint, owing to stronger weathering in the basement than in the overlying sandstones. No local tectonic control is identified along the eastern scarps that face the sedimentary wall (Serra das Matas, Serra das Pipocas).

The glint itself reaches its maximum heights in the north, above Freicheirinha and Mucambo (600–700 m), where short dissected concave slopes and pediments, only interrupted by a few quartzite ridges (Viçosa do Ceará) and scattered limestone hills, are shaped below the sandstone ledge. South of the Guaraciaba promontory, a sandstone body protruding to the north-east owing to its preservation in a narrow graben formed on the Transbrasiliano shear zone, the escarpment trends south, only interrupted by the consequent funnel of the Poti river (Fig. 3.2).

The lack of outliers underlines the relatively high dip angle of the sandstone layers toward the center of the Parnaiba basin (Fig. 3.5c, CD, EF, GH), and probably implies the former bevelling of its eastern edge together with the basement—e.g., the downwarped Guaraciaba compartment is also the place where the greatest thicknesses of sandstones are preserved, forming the only notable salient of the glint, and its highest point (1099 m; Fig. 3.2). However, in most places, the back-slope of the glint is an exhumed structural plateau, slightly tilted westward, and dissected by angular networks of fault- or fracture-line valleys. To the south, the escarpment becomes subdued, except around the heads of several tributaries of the upper Jaguaribe river, above depressions and ridges shaped by differential erosion along an EW exhumed molassic graben (Cococi; Fig. 3.2). Its southern tip lowers as the sandstone cover thins out: the sedimentary plateau hardly overlooks the exhumed infra-Paleozoic and pre-Cenomanian erosion surfaces capped by laterites (Quixariú-Campos Sales area).

3.1.3 The Borborema Plateau Region: Southeast Shoulder of the Cariri-Potiguar Rift Zone

The southeast shoulder of the Cariri-Potiguar rift zone lacks stratigraphic landmarks which might allow an evaluation of vertical movements or even the location of reactivated faults. Only the identification of pseudotachylites cutting across sedimentary units in the Rio do Peixe basins indicates exhumation by post-Cretaceous movements (thrusting of the basement) along the southern border of these basins (Nogueira et al. 2015). The main faults identified as belonging to the rift zone in this area are those of the Portalegre shear zone (along which the small Gangorra basin is downfaulted, north of Umarizal: Corsino and Tiriba 1985; De Castro et al. 2012), the Carnaubais fault system, which separates the 4000 m deep Potiguar rift from the Touros platform in the continuation of the Portalegre shear zone, and the Apodi transfer fault (Fig. 3.1). The other important structure which was reactivated during the Neocomian rifting, with NW throw, is the Rio Piranhas fault, at the southeast border of the Pombal basin.

The syn-rift bevelling of the Carnaubais and Apodi fault structures, their burying below the unconformable Cenomanian Açu sandstones and the lack of important parallel fault zone between them and the northernmost abrupts of the Borborema uplands (Serra do João do Vale, Serra de Santana; Fig. 3.5b, OP and c, KL) suggest the erosional origin of these scarps, located 20–30 km from the master faults (Fig. 3.2). Morphostratigraphic relationships also suggest that retreat of these fault scarps and corresponding development of the palaeopediment now exhumed from the post-rift cover had mostly occurred as soon as the Late Cretaceous (Peulvast et al. 2008).

Such evidence is lacking in the Pombal area, where no post-rift cover is identified. The presence of high inselbergs and ridges towering above the Sertaneja surface only a few km SE from the Rio Piranhas fault shows that scarp retreat remained irregular (Figs. 3.2 and 3.3). Moreover, the restricted dimensions of the tectonic basin (less than 20 km in length and 10 km in width), its asymmetry and the presence of highlands at similar altitudes (up to 700–800 m) on both sides, suggest that the topographic contrast created by rifting between the numerous horsts or tilted blocks of this part of the rift zone and its southeast shoulder was limited (Fig. 3.5c, KL). It may have been compensated during or after rifting by erosion and/or sedimentary infilling, before later dissection and exhumation.

The absence of Cretaceous sediments south of the Rio Piranhas fault system suggests that no important basin or embayment had developed into the southeast shoulder and that, together with the horsts which separate the Pombal and Pau dos Ferros basins, this area belongs to a wide block which was uplifted south of the Potiguar rift basin, and probably deformed during rifting by minor synthetic faults related to the Rio Piranhas and Patos fault systems (Fig. 3.2). Movements along the Cedro branch of this last shear zone may have raised the Borborema plateau to its highest altitudes as the upper edge of a tilted block (Serra do Texeira, 1197 m; Fig. 3.5b, ST). However, later differential erosion between resistant granites and softer metamorphic rocks of the Transverse Zone domain may also explain the presence of this granitic ridge between the low Patos basin (250 m) (Fig. 3.2) and the high Texeira surface (600–700 m) developed on its backslope.

Only a few scarps are controlled by fault lines, except for long EW granitic ridges which form the south limit of the Pombal depression on both sides of the Patos shear zone, south of Piancó (Fig. 3.1). Most escarpments appear as residual ridges between bands of sheared rocks, and present similar altitudes on both sides of the fault lines. However, the numerous parallel ridges which curve from WE to SSW–NNE around the Patos-Caicó lowlands are arranged in a kind of wide staircase which might correspond to complex faulting of the former rift shoulder. Erosional plains and embayments connected to the Sertaneja surface are carved into gneisses and migmatites of the rift shoulder, between the Borborema plateau and the Rio Piranhas fault system. Their width reaches 60 km. Around the Pombal basin, the planation surface is wider in the footwall (SE) than in the hanging wall of the Rio Piranhas fault (Fig. 3.5c, KL). On both sides of the large granitic inselgebirge of Serra Negra do Norte (up to 700 m), it is subdivided into two plains which widen around tributaries of the Piranha River flowing from the Borborema highlands (Caicó-Jardim de Seridó, Pombal). South of them, smaller embayments are developed into the Borborema plateau along crushed rocks of the Patos fault zone (Piancó and Patos depressions), between spectacular inselbergs and residual ridges (Fig. 3.2). Therefore, this area is so strongly shaped by dissection and differential erosion that tectonic patterns of the former rift can hardly be recognized.

Rolling remnants of high erosion surfaces (600–750 m) are identified on the Borborema plateau, forming a long WSW trending divide south of the granitic Teixeira ridge (Teixeira surface, Serra do Cedro), between the Rio Piranhas catchment and a wide hanging depression drained eastwards by the Paraiba and Pajeú rivers (the “Fossé médian du Nordeste”; Demangeot 1960). Other high residual ridges (up to 1170 m) overlook this surface at the border between the Paraiba and Pernambuco states. Westwards, similar plateaus form the divide between the Piancó and Milagres depressions (Monte Horebe, 800 m). Above Diamante, they locally derive from the pre-Paleozoic planation surface exhumed at the periphery of the Araripe basin (Fig. 3.5b, UV). Here, the westward dip of this unconformity surface reflects a flexural uplift of the Borborema plateau above the Cretaceous basin, but this dip also implies that all other landforms of the Borborema plateau, including the Teixeira surface, have been shaped by syn- and post-rift erosion at relatively deep levels below this old deformed surface.

To the northeast, the opposite extremity of the rift shoulder is much more dissected. A wide set of parallel crests of metamorphic rocks trending NNE is breached by rivers flowing from the wide upland (600 m) which forms the divide between the Rio Piranhas catchment, 30 km east of these ridges, and the east slopes of the Borborema (Fig. 3.2). Shallow hanging depressions are excavated into this plateau along a wide NS shear zone, around the upper reaches of these tributaries of the Piranhas River (Seridó, Picuí and Currais Novos depressions). Together with the ridge tops, the plateau progressively lowers northwards, until forming a digitated promontory—Serra do Feiticeiro, 500 m—which ends abruptly above the EW Açu-Lajes depression.

The ridges may result from dissection of an erosion surface, a remnant of which is preserved at 600 m above sea level in the Serra de Santana, below the Serra do Martins sandstones. The unconformable sandstone layers of this formation are hardly deformed (Menezes et al. 2002; Morais Neto and Alkmim 2001), even on the continuation of sigmoid faults branching on the Patos shear zone south of Seridó (Fig. 3.5b, OP and c, KL). Therefore either these faults are minor structures, non-reactivated during the Cretaceous rifting—in the case where the sediments are pre-Cretaceous in age, or the Serra do Martins sediments are remnants of a post-rift cover preserved on the bevelled rift shoulder and its faulted structures, apart from the identified Neocomian rifts. Whatever be the interpretation (see discussion in Sect. 5.3.2), it appears that differential vertical movements, if some, were concentrated on the Rio Piranhas fault, and that differential erosion (excavation of wide depressions, irregular scarp retreat into the rift shoulder, mainly at the expense of migmatites) is the main process involved in the present conformation of the Borborema scarp. Although very thin, the sandstone cover itself seems to have protected the basement from erosion, leaving the Serra de Santana as a big outlier between the Caicó and Lajes depressions (Fig. 3.3).

3.2 Drainage, Tectonics, and Differential Erosion

3.2.1 Regional Patterns

The regional drainage is mainly oriented northeast, normally to the general trend of the coastline. This orientation underlines the direction in which the Jaguaribe-Piranhas amphitheater opens on the equatorial Atlantic Ocean (Fig. 3.2). However, since no centripetal or annular pattern (Deffontaines and Chorowicz 1991) fully reflects this topography, it can be assumed that the amphitheater pattern does not predate the drainage installation or that mechanisms of epigeny or capture were able to overcome the structural controls involved in its development (Peulvast and Vanney 2001).

The westernmost part of the amphitheater is drained by *a centrifugal river*, the Rio Poti, which flows westwards from the Central Ceará highlands through the Crateus depression, the Ibiapaba glint and a cataclinal funnel, toward the Parnaiba basin. It may be a consequent river originally settled on the basement and its sedimentary cover, both bevelled by a westward-sloping planation surface (Saadi and Torquato 1992) on the back-slope of the northwest shoulder of the Cariri-Potiguar rift zone (Fig. 3.5c, EF, GH). Later downcutting resulted in the excavation of the Crateus depression into the basement and in the shaping of the glint.

The catchments of all other rivers fully belong to the amphitheater, which is bounded by a drainage divide mainly coinciding with the topographic crest. Exceptions are only recorded in the northern Borborema plateau, where the upper reaches of a few tributaries of the Piranhas River widen behind high ridges at the edge of the rift shoulder above the Sertaneja surface. Even in this case, no hook pattern is identified that might suggest the occurrence of captures at the expense of the back-slope of the massif (Fig. 3.14). This statement remains true along the south crest of the amphitheater, in spite of the northward bend of the upper Jardim River, first oriented to SE in the southeast part of the Chapada do Araripe, and of much deeper erosion at shorter distances from the crest than in the nearby São Francisco catchment.

The main interior divide corresponds to the Central Ceará highlands, i.e., the northwest shoulder of the Cariri-Potiguar rift zone. To the north, between the Choró and Coreaú valleys, narrow catchments with parallel or diverging rivers only form 27 % of the total area drained toward the continental margin of the studied region. The remaining area (73 %) is drained by the Jaguaribe and Piranhas rivers, and by shorter rivers converging toward the Potiguar basin. Mainly trending NE, this drainage system reflects the horst and graben pattern of the Cariri-Potiguar rift zone, following fault lines, grabens, and other faulted structures, but also comprises orthogonal segments following the back- and front slopes of tilted blocks and escarpments (Fig. 3.14). In both cases, the main divides are located 100–200 km inland of the marginal escarpment, whereas shorter rivers reflect the establishment of a secondary divide at the escarpment (or its innermost reliefs).

This drainage pattern strongly underlines the Cretaceous tectonic heritage, since the main rivers either follow Neocomian rift structures or flow toward them from their shoulders, like in Leeder and Gawthorpe's models (1987). It may explain the strong asymmetry of the mid Jaguaribe basin, where the main river trunk, downstream of Orós, only receives a short tributary on its right (east) side (Rio Figueiredo), whereas it receives on the opposite side long rivers draining the northwest shoulder of the Potiguar rift, reflecting a former hanging/foot wall slope opposition related to the Jaguaribe fault system (Fig. 3.14). Nevertheless, this pattern may also reflect an attraction related to post-rift subsidence of the Potiguar basin, as suggested by the trend of the Banabuiú river, which turns east and leaves the shallow corridor excavated along the Senador Pompeu shear zone, crossing the Orós-Serra do Felix ridge rather than continuing NE toward the Piranji river—the only

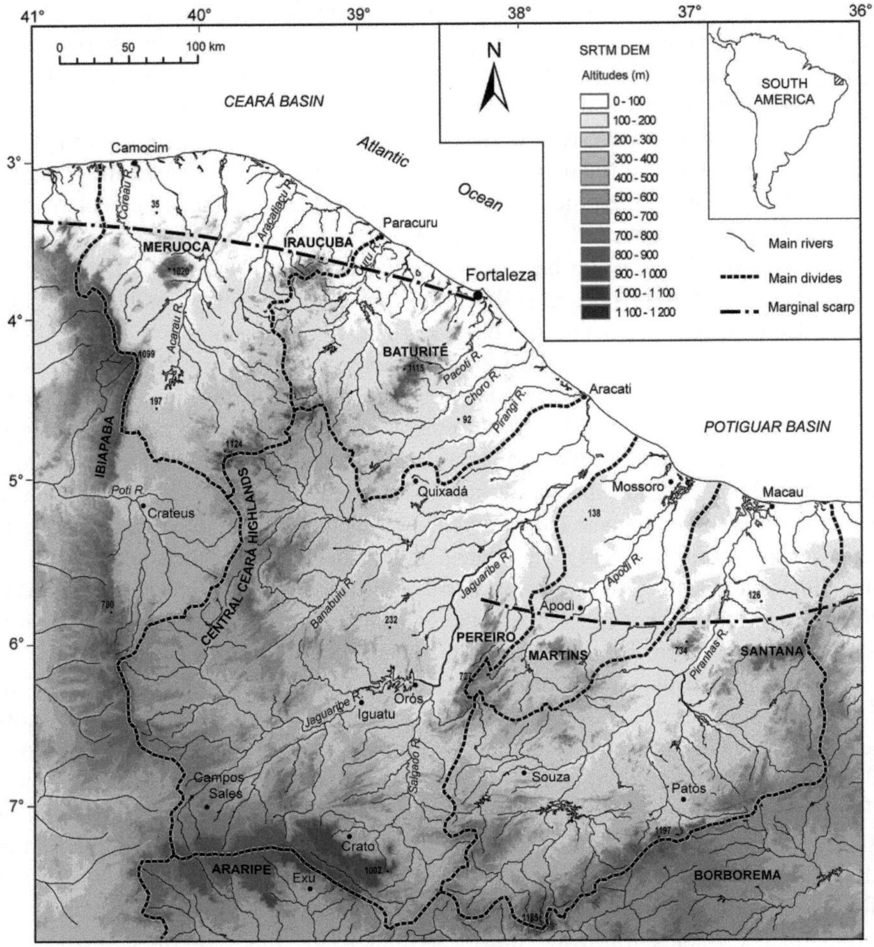

Fig. 3.14 Topography and the main drainage systems in the northern Brazilian Nordeste

important river in Ceará which is not fed by highland sources. It may have been captured by tributaries of the lower Jaguaribe river—the present lower Banabuiú and Sitia rivers, more easily deepened owing to the regional dip toward the Potiguar basin and to the weak resistance of its sediments to erosion (Fig. 3.5b, OP).

3.2.2 Structurally Controlled River Systems

Most *short rivers of the oceanic front* follow the regional slope of the Sertaneja surface and of the coastal sedimentary plateaus toward the ocean. They flow from the marginal escarpment, locally between residual ridges and inselbergs protruding through the tabuleiros (Coreaú and Choró rivers; Fig. 3.3). Originating in gullies and cirques abruptly incised into these mountains (Irauçuba and Baturité massifs) or in remote inselgebirge and inselberg groups (Quixadá-Serra Azul: Choró and Piranji rivers), their wide valleys slightly incised into the low surface hardly deepen when crossing the tabuleiros. Their lower reaches enter these low plateaus through wide funnels where the basement underlies the alluvial floor down to the coastal area (Choró river), between faint sedimentary slopes. The easternmost rivers flow through the wider and higher Chapada do Apodi, east of the Jaguaribe river, widening in the orthoclinal depression of Apodi-Açu-Lajes before crossing the low cuesta and finally narrowing between slopes of decreasing height (Apodi (Mossoró) and Piranhas (Açu) rivers).

Only a few rivers of this category are controlled by fault lines (Choró river upstream of Pacajus, and Piranji river upstream of Cristais, along faults of the Senador Pompeu shear zone; Fig. 3.2). Within the Potiguar basin, the lower Mossoró and Açu valleys clearly follow the NE-trending hinge lines on both sides of the Serra do Mel anticline, reflecting an active tectonic control along reactivated fault lines (Maia and Bezerra 2014b).

A few rivers are *guided by outcrops of soft rocks*. The most typical of them drain subsequent depressions below cuestas and glints. The lower Jaguaribe valley between Tabuleiro do Norte and Itaiçaba is a wide asymmetric depression shaped along the contact between the basement and the Açu sandstone, below the cuesta of the Chapada do Apodi (Fig. 3.15). Together with the regional dip of the strata

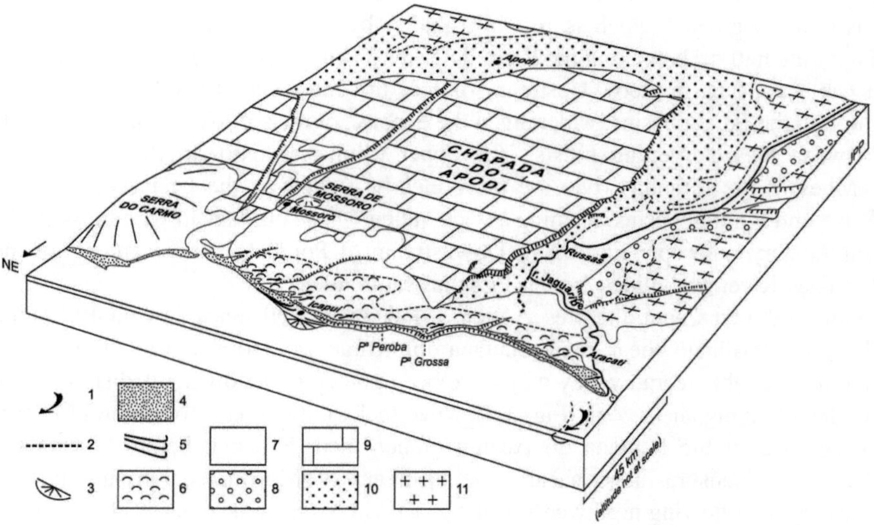

Fig. 3.15 Sketch block-diagram of the Apodi-Aracati area. *1* former channel of the Jaguaribe river; *2* limit of geological unit; *3* ebb tidal delta; *4* active dunes; *5* sand barriers; *6* palaeodunes; *7* Barreiras sediments; *8* Faceira Formation (fluvial terrace); *9* Jandaira limestone (Turonian-Campanian); *10* Açu sandstone (Cenomanian); *11* basement

toward the center of the Potiguar basin, which controls the N to NE trend of the sedimentary outcrops, this lithological control overcomes the control by the N20 to N30 Jaguaribe fault lines which guide the river upstream of Jaguaribara (Fig. 3.2). The coincidence between the regional slope toward the ocean and the trend of the contact zone probably helped the river to arrange this pattern, whereas nearby rivers were only able to form epigenic valleys through the chapada. A few rivers follow tectonically similar depressions below the Ibiapaba glint, but their valleys are excavated into the basement: the Groaíras River east of Ipu and the upper Poti river, south of Crateus. Around the eastern tip of the Chapada do Araripe, the Crato-Milagres sedimentary depression is also drained by roughly longitudinal rivers (upper tributaries of the Salgado River) and its easy excavation probably caused the northward bend of the upper Jardim River to the Salgado system.

More local or simpler structural controls correspond to the excavation of narrow bands of sheared rocks between outcrops of granites or gneisses, by longitudinal river segments connecting oblique or transverse segments. Fault line valleys are identified north of the Chapada do Araripe along the Tatajuba and Potengi shear zones (Rio dos Bastiões along the Tarrafas graben), and along the Senador Pompeu shear zone (the straight upper Banabuiú River; Fig. 3.2). They are widespread, at smaller scales, in all the highlands, where they underline complex, often orthogonal (Meruoca, Baturité massifs) networks of faults, fractures, and shear zones inherited from the tectonic history of the basement.

Several important rivers drain *wide tectonic depressions*, mainly half-grabens formed during the Cretaceous rifting but subsequently widened by erosion and only partly underlain by sediments. The main of them is the lower Salgado-mid-Jaguaribe segment, which is roughly guided by the fault angle of the complex Jaguaribe half graben along the Serra do Pereiro (Fig. 3.2), but only locally follows a Cretaceous basin (Icó). Upstream of Icó, the Jaguaribe River flows obliquely through the Iguatu basins, widening at the expense of the soft Neocomian sediments between the intervening horsts. The upper tributaries (Rio do Peixe) and mid reaches of the Piranhas river are more clearly guided by the Cretaceous Rio do Peixe and Pombal basins, forming a wide valley whose outlines fit with those of the sedimentary outcrops (Fig. 3.8a). Downstream of Pombal, some segments of the Piranhas River are guided by the Piranhas fault zone.

In northwest Ceará, the preservation of Paleozoic sandstones downfaulted along the Transbrasiliano shear zone (Santana do Acaraú graben) also strongly suggests that the straight Acaraú valley may have been guided by a downfaulted structure or by the adjoining shear zone (Fig. 3.3). Nevertheless, the river only follows the NW shear zone of the Santana do Acarau graben, near Santana, before leaving the excavated Transbrasiliano shear zone upstream of Morrinhos, through its NW shoulder, and flowing northwards to the coast without tectonic control. As discussed higher, a tectonic control may also be suggested for parts of the Canindé and Curu rivers, but cannot be demonstrated because of the lack of pre-Neogene sediments.

3.2.3 Epigeny Phenomena and Their Morphological Meaning

Most other rivers of the study area are more independent relatively to the structure: they cut through fold and fault structures of the basement and residual ridges shaped into the hardest rocks (quartzites, granites). Among the corresponding water gaps or *boqueirões*, several are dammed, and now retain large lakes (Açudes Banabuiú, Riacho do Sangue, Orós, Piranhas, Mãe d'Agua; Fig. 3.2).

Two main groups of transverse rivers and water gaps are identified: the N to NW-trending streams that flow from the Borborema highlands to the Jaguaribe and Piranhas plains through ridges and corridors of the Patos and Seridó shear zones (Cariús river and smaller creeks of the Serra do Quincuncá; Salgado water gap: 93 m deep, 40 m wide; Piancó, Espinharas and Seridó rivers), and the WE valleys that drain the western Jaguaribe plain from the Central Ceará highlands to the Jaguaribe river, through the quartzitic Orós-Banabuiú ridges (Jaguaribe, das Pedras and Banabuiú Rivers). North of Banabuiú, the same ridges form a water divide on the Aracati platform although becoming much lower (Fig. 3.3). Both systems of ridges, as well as the granitic ridge which encloses the small Icózinho basin at the south tip of the Pereiro massif, are crossed by transverse rivers only in places where they separate hanging wall or foot-wall ridges from fault angle valleys shaped along Neocomian rift structures. This distribution may be a convincing clue for an epigenic origin, e.g., superimposition and incision from a former post-rift erosion or filling surface which may have been linked to the Upper Cretaceous sedimentary surfaces of the Araripe and Apodi plateaus (Fig. 3.5b, QR, ST and c, IJ).

Other water gaps are only found in the south part of the Central Ceará highlands (Saadi and Torquato 1992), where they were cut through low and discontinuous ridges by the upper Jaguaribe river (Arneiroz) and by shorter SE trending rivers draining the north-west rift shoulder toward the plain, around Acopiara and Mombaça (Truçu and upper Banabuiú rivers). Here, the altitudes of the crests are consistent with the same hypothesis (Fig. 3.2). On the contrary, the Piranhas gap, shaped between the Serras of Santana and João do Vale (Jucurutu), is better explained by widening of faulted structures in the continuation of the Piranhas fault zone.

The distribution of epigenic water gaps suggests that superimposition phenomena occurred mainly in the southwest and south highlands and depressions of the Borborema-Ibiapaba hemicycle, i.e., around the highest remnants of non-deformed pre-rift cover (southern Ibiapaba) and of the post-rift sedimentary cover (Araripe), and through ridges of similar or lower altitudes. This fact may indicate that at least one major infilling or planation event of the former rift structures have played a role (Peulvast and Claudino Sales 2004; Peulvast et al. 2008). Epigeny phenomena may also be suspected in northwest Ceará (Rio Coreaú and tributaries), where several rivers flow through EW quartzitic ridges shaped by differential erosion into the folded and once bevelled basement, north of the Ibiapaba glint (Fig. 3.2). Nevertheless, river head retreat may have occurred during the evolution of this well-defined part of the marginal escarpment, as well as superimposition from an erosion surface bended and downwarped by the flexural movement of the margin (Peulvast et al. 2008).

The morphostructural study of the Chapada do Araripe brings more indications on the epigeny process and its participation in basin inversion (Peulvast and Bétard 2015). Beyond the general westward inclination of this plateau, gentle slopes are observed toward the axis of the shallow valley systems oriented to the SE, and also around the heads of much shorter catchments which hang over a few scarp re-entrants and box-canyons of the north side (Fig. 3.6). This hanging dissection topography (absent in the east, over the Cariri depression, where the scarp directly cuts the highest parts of the plateau) suggests that shallow fluvial erosion has taken place on the sandstone cover, independently of (or earlier than) the topographic inversion processes, restrained to the periphery of the plateau.

Except for the Araripina valleys, this drainage system is dry and disconnected from that of the surrounding depressions. The system is independent from the general westward dip of the sandstone cap and from the underlying NE–SW to EW rift structures. Its configuration and gently sloping longitudinal profiles probably reflect the early coincidence between the EW axis of the basin and a water divide located close to the north rim of the plateau, in times when the relief of the Chapada already was culminant, forming a low and weakly dissected ridge. The present lack of superficial flow in these valleys, as well as the permeability of the sandstone cap also imply an elevated level of the water table at the time of their formation and, therefore, a weak relief and shallow dissection of the surroundings, with shallow regional base levels. As suggested by the lack of hanging valleys in the northeast, the water divide only migrated in this sector during the later and deeper dissection stages, because of a strong retreat of the heads of the Salgado river system into the plateau. In other cases, the early and shallow dissection prepared the deepening stage, leading to the formation of dendritic systems of canyons inside the plateau (Araripina).

3.2.4 The Lithological Filter

Our study stresses the decisive role of differential erosion in the landscapes of the study area, in spite of many uncertainties on the persistence of residual tectonic landforms or on a possible role of neotectonics (Pereiro massif: Gurgel et al. 2013). However, the available lithological data only give a part of the numerous parameters involved in the resistance of rocks to selective erosion (Lagasquie et al. 2001; Peulvast and Vanney 2001). Many landforms of differential erosion are described in interior regions, in the basement and at the contact between basement and sedimentary covers, in places where direct tectonic control on landforms can be excluded (dissected erosion or structural surfaces, for instance). They are confirmed by ^{10}Be analyses on rock outcrops within the Sertaneja depression (Seridó valley, RN), which show moderate (4–11 m/Myear) erosion rates, with lower rates (<3 m/

Myear) on quartzite domes and slightly higher rates on micaschists (3–7 m/Myear) and pegmatites (8–13 m/Myear) (Morais Neto et al. 2010).

Lithological controls are less clearly defined in the outlines of the main mountain volumes overlooking the pediplains of the Sertaneja surface (Fig. 3.2). Their signification remains ambiguous because of numerous cases of escarpments abutting on fault lines possibly reactivated in the Cretaceous or later (granitic plateaus of Pereiro and Meruoca: Peulvast et al. 2006; Peulvast and Bétard 2013; see also Sect. 5.2). In spite of these ambiguities, a few general observations are reported below.

3.2.4.1 Differential Erosion in Uplifted Basement Areas

Within uplifted basement areas, the most frequent type of selective erosion corresponds to incision and widening of depressions along fault lines and shear zones. It is observed in all the geological domains which form the Borborema province, and along the shear zones which separate these domains (Vauchez et al. 1995; Maia and Bezerra 2014a).

The petrographical heterogeneity of the basement is also frequently reflected in the landscape. Many contacts between syn- or late-Brasiliana intrusions and the metamorphic host rock control with more or less accuracy, but not systematically, the outlines of inselbergs, inselgebirge or wider plateaus. Among the mesozonal synorogenic granites, the Pereiro (571 ± 2 Myear: Magini and Hackspacher 2005) and Quincunca granites form prominent plateaus between gneissic corridors and ridges, whereas others are partly or totally excavated together with the migmatitic or gneissic host rocks, in the Seridó belt or in the Crateus depression (Fig. 3.3). The granitic cores of migmatitic nappes of the Central Ceará domain form some of the highest mountains of this area (north Baturité and Irauçuba mountains; Fig. 3.5b, MN; Brandão 1995; Bétard 2007). The higher level, late- or post-orogenic intrusions, classically considered as more resistant because less tectonized than older intrusions and host rocks (Birot 1981), more systematically form salient landforms, huge inselgebirge (Serra das Matas, Serra Azul) or big inselbergs (Serra da Barriga).

Around Quixadá, more detailed petrographical data also point out a role of the mineralogical composition. The Quixadá-Quixeramobim granitic complex is partly truncated by the Sertaneja surface (Fig. 3.10). In the calc-alkaline granitic complex, bornhardts and spectacular inselbergs are mainly preserved in the leucocratic facies of a late or post-orogenic shoshonitic suite (monzonites, biotite-leucogranites: Almeida et al. 1999). On the contrary, most intermediate and basic facies, more easily weathered (Godard et al. 2001; Peulvast and Vanney 2001), are bevelled, although a swarm of spectacular but relatively small bornhardts protrude from the flat surface in the less fractured areas (diorites and quartz-diorites, granodiorites).

The morphological expression of the Meruoca and Mucambo intrusions (syenogranites, monzonitic granites) is more ambiguous, because of the possible role of recent tectonics along one of the fault lines (the Café-Ipueiras fault) that form their limits (Peulvast et al. 2006).

Strong resistance contrasts are also observed within the metamorphic units. In the Central Ceará, Orós-Jaguaribe, and Rio Piranhas domains as well as in the Seridó belt, the migmatites of the pre-Brasiliano basement are often excavated or bevelled, whereas gneisses and other metamorphic rocks belonging to autochtonous parts (Pedra Branca) or thrusted slices of this basement and to the Proterozoic cover sequences more frequently form ridge complexes, plateaus, or higher mountains (Baturité). Among them, quartzite belts and marble lenses play a particular role. Synclinal bands of quartzites form long residual ridges whose arcuate outlines underline those of folds and major structural contacts, on the coastal surface south of Granja, or on the floor of the Sertaneja pediplain from Iguatu to the Piranji river (the Banabuiú-Orós and Serra do Felix ridge; Fig. 3.11). Marble lenses included in the supracrustal series may be recognized by their prominent and jagged landforms, in spite of their strong karstification (giant karren, caves). They form buttes and promontories below the Ibiapaba glint (Ubajara) and the Serra do Martins, or high crests in the eastern flank of the Baturité massif (Redenção). In several places, especially below the Paleozoic paleosurface, deep weathering of other metamorphic rocks largely prepared a selective removal of the surrounding material, resulting in the emergence of marble buttresses.

More complex situations are also found. Lithological controls involved in the excavation of the large Crateus depression are only partly deciphered. The resistance contrast between the Paleozoic sandstones and their weathered basement is the main factor explaining its location and asymmetry (Fig. 3.14), but the other outlines of this depression are less clearly controlled by lithological contacts, except for the granitic hills that almost close it to the north (Fig. 3.3). The deepest part of its floor is excavated in gneisses, but it also extends to the east and the south in migmatites and syntectonic granites, around a NS shear zone forming the axis of the depression (the Tauá fault zone). Therefore, we suspect here a large set of lithological controls. A good illustration of this complexity is provided by the landscapes of the Appalachian-like valley-and-ridge system shaped around NE-trending fault zones branching on the Patos shear zone, north of the Chapada do Araripe (Fig. 3.6). Here, many straight, sinuous or hairpin bevelled crests and narrow corridors reflect differential erosion along alternating soft and hard layers of the folded metamorphic basement, whereas wider ridges are formed by selective exhumation of elongate synkinematic granitic intrusions (around Varzea Alegre).

3.2.4.2 Contact Between Crystalline Basement and Sedimentary Cover

The contacts between basement and cover rocks are systematically underlined by differential erosion, localizing some of the highest peripheral scarps of the Jaguaribe-Piranhas hemicycle such as the Ibiapaba glint. As shown around the Chapada do Araripe (Peulvast and Bétard 2015), they form either glints, where hard rocks (sandstones of Paleozoic, Cenomanian, or uncertain age) directly overly the weathered basement submitted to stripping and excavation when exhumed, or cuestas where soft layers of the transitional and post-rift series separate the basement from the hard top layers (sandstones or limestones). In this last case, the excavation of wide lowlands (Crato-Milagres; Apodi-Lajes) is accompanied by exhumation of the basement, locally preserved with its Paleozoic sandstone tegument (Crato-Milagres depression), owing to the high resistance contrast between these deposits and the basement rocks, generally sound below the post-rift series. Opposite to the sedimentary scarps, these depressions are frequently surrounded by fault-controlled scarps (Fig. 3.5).

3.2.4.3 Lithological Controls and Stepped Topography

Only a few escarpments overlooking the low surfaces are clearly controlled by lithological contacts or tectonic structures. It could not be determined in all cases whether they are residual or even active fault scarps, or fault-line scarps, formed by differential erosion in relation with cyclic development of low planation surfaces (Peulvast et al. 2006). In the semi-arid conditions which prevail in interior lowlands, these escarpments and other structural landforms are generally well developed, with overall concave profiles or sharp breaks in slope. Their origin will be discussed in relation with that of stepped landfoms often reflecting lithological controls on the landscapes (see Sect. 4.1), since any cyclic evolution is known to have a better morphological expression if the lithological potential is contrasted enough to preserve the different storeys (Gunnell 1996; Peulvast and Vanney 2002).

3.3 The Determining Influence of Superposed Structural Heritages

The present relief of the study area reflects the juxtaposition of contrasted elements of varied ages, and a complex history. Many features are inherited from the most significant events which left their imprint in the geological structure of the study area: the Brasiliano orogeny, the late-orogenic collapse leading to the formation of the Parnaiba basin to the west (Teixeira et al. 2004; De Castro et al. 2014), the erosion and partial burying of the mountain range, the Neocomian intracontinental rifting, aborted and followed in the Albian by post-rift subsidence of former rift structures and transform oceanic opening, and at last the formation of the passive margin. This pattern implies a differentiated tectonic evolution of compartments largely inherited from the rifting and breakup patterns (Peulvast et al. 2008). In spite of the predominance of erosion regimes since the end of the post-rift stage, and of a significant role of differential erosion, there is a relatively good agreement between the regional hemicycle pattern and the effects of the successive tectonic events described here. Among the structural heritages, the numerous shear zones formed at depth in ductile regime during the Brasiliano orogeny play one of the most important roles, as well by focusing the later brittle deformations when conveniently oriented with regard to the stress fields (Maia and Bezerra 2014b), as by forming weakness zones when exposed to erosion.

The central depression of the hemicycle only partly corresponds to the former Cariri-Potiguar rift, since the southwest extremity of this structure has been uplifted together with the surrounding basement after the end of the post-rift stage. Most fault zones having guided the Early Cretaceous rifting correspond to reactivated Late Brasiliano shear zones (Matos 1992). Some of them still control residual fault scarps, whereas other shear zones, reactivated or not, control major features of differential erosion, fault line scarps and fault line valleys. Therefore, the regional morphology is largely controlled by these E- to NE-trending structures, mainly in the inner parts of the study area, but not only (for instance, the Transbrasiliano fault zone controls one of the major indentations of the oceanic façade of western Ceará).

In fact, the main tectonic controls on the large-scale features of the regional relief may probably be found in the patterns created during the post-rift and oceanic opening stages rather than in the continental rift structures themselves, which control smaller and more contrasted units such as rift shoulders, basins and intervening horsts and fault scarps. To the post-rift Potiguar synform correspond the main re-entrant in the oceanic façade, the offset zone between two highly contrasting parts of the marginal escarpment, and the center of the Jaguaribe-Piranhas hemicycle. Intersected by the transform zone of oceanic opening, this structure is both the result of thermal post-rift detumescence of the Potiguar rift and of the subsidence of the passive margin (Mello 1989; Matos 2000). Later on, this area, which we propose to call "Jaguaribe-Potiguar compartment," underwent a wide and gentle flexural movement which allowed the preservation of palaeosurfaces of Cretaceous age in the hinge line, at the foot of residual fault scarps (Peulvast et al. 2008).

The opposite compartment—the "Baturité-Irauçuba compartment"—is represented by higher summits near the coast, reflecting a more vigorous uplift above the Aracati platform and the Ceará basin. On the basis of a comparison with the preservation of Cretaceous palaeo-escarpments south of the Potiguar basin, it was suggested that this contrast (materialized by the eastern escarpment of the Central Ceará mountains) might be a direct heritage of differential shoulder uplift during the syn-rift, post-rift, and oceanic opening stages, but such an interpretation does not preclude the role of stronger uplift in later times, related to steeper flexural deformation of the margin west of Fortaleza. In this case, decoupling might have occurred along the Senador Pompeu fault zone (Peulvast et al. 2008).

The outlines of our "Jaguaribe-Potiguar compartment" geographically fit with those of a 250 km wide crustal thinning zone more or less coincident with the Potiguar basin (De Castro et al. 1997, 2012; Fig. 3.1). Interfering with the regional seaward crustal thinning (from 30–31 km inland to 15 km below the continental rise), it reaches 2–3 km relatively to the adjacent areas, and its SSW-trending axis corresponds to extension along the ESE-WNW direction which is that of the Neocomian-Barremian continental rifting. On this basis, Barros et al. (1999) suggested that the present topography of the study area reflects the crustal response to the Early Cretaceous extensional deformation. On the contrary, a maximum thickness zone (more than 31 km) of NNE–SSE direction is recognized in western

Ceará, below the Serra da Ibiapaba, that is to say, in coincidence with the western edge of the Jaguaribe-Piranhas hemicycle. According to De Castro et al. (1997), it would represent a former Brasiliano suture zone coinciding with the Transbrasiliano shear zone. To the east and southeast, the Borborema plateau also coincides with a zone of normal crust thickness.

Areas where extension and crustal thinning were more limited only underwent transient and geographically limited subsidence, as demonstrated by the evolution of the uplifted Araripe basin, below which the crustal thickness is about 30 km. The subsidence of the Potiguar basin—which in fact affected the whole Jaguaribe-Potiguar compartment because of the mainly flexural behavior of the continental lithosphere—was enhanced by that of the continental margin which intersects the former rift basin (Mello 1989). That such a subsidence is better explained by crustal thinning of the margin and post-opening thermal evolution of the underlying lithosphere ("tectonic subsidence": Boillot and Coulon 1998), rather than by the sedimentary load, remains a matter of discussion (Bishop and Goldrick 2000; Peulvast et al. 2008).

The western limit of the crustal thinning zone approximately coincides with the Senador Pompeu fault zone (Fig. 3.1). This deep weakness zone had separated the zone of maximum crustal thickening and faster cooling (to the north) from that of lesser thickening, upright folding, high-temperature, low-pressure conditions, and slower cooling (to the south) during the Brasiliano orogeny (Monié et al. 1997). It may have focused local deformations and major decoupling between adjacent crustal compartments instead of the flexural movements responding elsewhere to the isostatic compensation. Beyond the seismic activity recorded along this fault, it may explain the maintenance of the eastern escarpment of the Central Ceará highlands, especially if additional causes of uplift, such as the Tertiary magmatism have enhanced the uplift of the western block. There is also an approximate correspondence between crustal isopachs and the eastern marginal escarpment, south of the Potiguar basin, but unlike Jardim de Sá et al. (1999), we did not find any evidence of localized fault or flexural movement which might explain the EW alignment of the Serra do Portalegre-Serra do Martins-Serra de Santana (Fig. 3.3).

References

Almeida AR, Ulbrich HHG, McReath I (1999) O batólito Quixadá – petrologia e geoquímica. Rev Geologia 12:29–52

Barros SDS, Castro DL, Matos RMD, Jardim De Sa EF, Saadi A (1999) The orography of northeast Brazil and the crust-mantle interface relief: the correlation between geomorphologic and gravimetric data. In: Abstracts of the Regional Conf on Geomorph, Rio de Janeiro, July 1999

Bétard F (2007) Montagnes humides au cœur du Nordeste brésilien semi-aride: le cas du massif de Baturité (Ceará). Apports d'une approche morphopédologique pour la connaissance et la gestion des milieux. Ph.D thesis, Univ Paris-Sorbonne

Bétard F (2012) Spatial variations of soil weathering processes in a tropical mountain environment: the Baturité massif and its piedmont (Ceará, NE Brazil). Catena 93(18–28):2012. doi:10.1016/j.catena.2012.01.013

Bezerra FHR, do Nascimento AF, Ferreira JM, Nogueira FC, Fuck RA, Brito Neves BB, Sousa MOL (2011) Review of active faults in the Borborema Province, Intraplate South America Integration of seismological and paleoseismological data. Tectonophysics 510:269–290

Birot P (1981) Les processus d'érosion à la surface des continents. Masson, Paris

Bishop P, Goldrick G (2000) Geomorphological evolution of the east Australian continental margin. In: Summerfield MA (ed) Geomorphology and global tectonics. Wiley, Chichester, pp 225–253

Boillot G, Coulon C (1998) La déchirure continentale et l'ouverture océanique. Géologie des marges passives. Gordon and Breach Sci Publ, Amsterdam

Brandão RL (1995) Mapa geológico da região metropolitana de Fortaleza. Texto explicativo. Sistema de informações para gestão et administração territorial da Região metropolitana de Fortaleza – Projeto SINFOR, CPRM

Corsino AR, Tiriba VF (1985) Nota sobre a occorrência de uma pequena bacia sedimentar ao norte do município de Umarizal (R.N) – (Bacia de Gangorra). Bol DG-CCE/UFRN, Natal 9:54–56

De Castro DL, Barbosa VCF, Silva JBC, Medeiros WE (1997) Relevo da interface crosta-manto entre os vínculos de isostasia e suavidade. Congr Int SBGf 5, São Paulo, Anais, pp 682–685

De Castro DL, Bezerra FHR, Sousa MOL, Fuck RA (2012) Influence of Neoproterozoic tectonic fabric on the origin of the Potiguar Basin, northeastern Brazil and its link with West Africa based on gravity and magnetic data. J Geodyn 54:29–42

De Castro DL, Fuck RA, Phillips JD, Vidotti RM, Bezerra FH, Dantas EL (2014) Crustal structure beneath the Paleozoic Parnaíba Basin revealed by airborne gravity and magnetic data, Brazil. Tectonophysics 614:128–145

Deffontaines B, Chorowicz J (1991) Principles of drainage basin analysis from multisource data: application to the structural analysis of the Zaire basin. Tectonophysics 194:237–263

Demangeot J (1960) Essai sur le relief du Nord-Est brésilien. Ann Géogr 69(372):157–176

Destro N, Szatmari P, Ladeira EA (1994) Post-Devonian transpressional reactivation of a Proterozoic ductile shear zone in Ceará, NE Brazil. J Struct Geol 16(1):35–45

Galvão CC, Da Silva FCA (2002) Análise microestrutural de um corpo siliciclástico localizado a Nordeste de Santana do Acaraú – CE. XLI Congr Bras Geol, João Pessoa, Anais, p 624

Godard A, Lagasquie JJ, Lageat Y (2001) Basement regions. Springer, Berlin

Gunnell Y (1996) Géodynamique d'une moyenne montagne tropicale. La genèse des paysages dans le Ghat occidental du Deccan, sur son revers continental et son piémont maritime. Ph.D thesis, Univ. B. Pascal, Clermont-Ferrand

Gurgel SPP, Bezerra FHR, Corréa ACB, Marques FO, Maia RP (2013) Cenozoic uplift and erosion of structural landforms in NE Brazil. Geomorphology 186:68–84

Jardim de Sá EF, Matos RMD, Morais Neto JM, Saadi A, Pessoa Neto OC (1999) Epirogenia Cenozózoica na Província Borborema: síntese e discussão sobre os modelos de deformação associados. VII Simp. Nac. Estudos Tect., Lençoís – Bahia, Anais, pp 58–61

Lagasquie JJ, Lageat Y, Peulvast JP, Godard A, Gunnell Y (2001) Morphostructural units, multi-storeyed scenery and the origin of escarpments in basement terrains. In: Godard A, Lagasquie JJ, Lageat Y (eds) Basement regions. Springer, Berlin, pp 35–63

Leeder MR, Gawthorpe RL (1987) Sedimentary models for extensional tilt-block/half-graben basins. In: Coward MP, Dewey JF, Hancock PL (eds) Continental extensional tectonics, Geol Soc Spec Publ vol 28. pp 139–152

Magini C, Hackspacher PC (2005) Evolução metamórfica de arcos magmáticos neo proterozóicos: região NE de província Borborema. Revista de Geologia 18(2):187–202

Maia RP, Bezerra FHR (2014a) Tópicos de geomorfologia estrutural, Nordeste brasileiro. Edições UFC, Fortaleza

Maia RP, Bezerra FHR (2014b) Inversão neotectônica do relevo na Bacia Potiguar, Nordeste do Brasil. Revista Brasileira de Geomorfologia 15(1):61–74

Matos RMD (1992) The Northeast Brazilian rift system. Tectonics 11(4):766–791

Matos RMD (2000) Tectonic evolution of the Equatorial South Atlantic. In: Mohriak W, Taiwani M (eds) Atlantic Rifts and continental margins. Geophys Monograph, vol 115. Am Geophys Union, pp 331–354

Mello UT (1989) Controles tectônicos na estratigrafia da Bacia Potiguar: uma integração de modelos geodinâmicos. Bol Geoci Petrobras 3(4):347–364

Menezes MRF, Lima Filho FP, Farias PRC (2002) Fácies e sistema deposicional da Formação Serra do Martins nos platôs terciários do Rio Grande do Norte. XLI Congr Bras Geol, Joã Pessoa, Anais, p 387

Monié P, Caby R, Arthaud M (1997) The Neoproterozoic Brasiliano orogeny in northeast Brazil: $^{40}Ar/^{39}Ar$ and petrostructural data from Ceará. Precambrian Res 81:241–264

Morais Neto JM, Alkmim FF (2001) A deformação das coberturas terciárias do Planalto da Borborema (PB-RN) e seu significado tectônico. Rev Bras Geociências 31(1):95–106

Morais Neto JM, Vasconcelos PM, Stone J, Lima MG (2010) Variability of erosion rates captured by ^{10}Be cosmogenic isotope: a case study in the Seridó valley, northeastern Brazil. VII SSAGI-South American Symposium on Isotope Geology, Brasilia

Nogueira FCC, Marques FO, Bezerra FHR, De Castro DL, Fuck RA (2015) Cretaceous intracontinental rifting and post-rift inversion in NE Brazil: insights from the Rio do Peixe basin. Tectonophysics 644–645:92–107. doi:10.1016/j.tecto.2014.12.016

Peulvast JP, Bétard F (2013) Late Cenozoic and present-day hillslope erosion dynamics in a passive margin context: stability or instability? Case studies in Northeast Brazil. Geografia Fisica e Dinamica Quaternaria 36(1):139–149. doi:10.4461/GFDQ.2013.36.0

Peulvast JP, Bétard F (2015) A history of basin inversion, scarp retreat and shallow denudation: the Araripe Basin as a keystone for understanding long-term landscape evolution in NE Brazil. Geomorphology 233:20–40. doi:10.1016/j.geomorph.2014.10.009

Peulvast JP, Claudino Sales V (2004) Stepped surfaces and palaeolandforms in the northern Brazilian "Nordeste": constraints on models of morphotectonic evolution. Geomorphology 62:89–122

Peulvast JP, Vanney JR (2001) Géomorphologie structurale, tome 1. Relief et structure. Gordon & Breach, Paris, and BRGM Editions, Orléans

Peulvast JP, Vanney JR (2002) Géomorphologie structurale, tome 2. Relief et géodynamique. CPI-GB Science Publ, Paris, and BRGM Editions, Orléans

Peulvast JP, Claudino Sales V, Bezerra FHR, Bétard F (2006) Landforms and neotectonics in the equatorial passive margin of Brazil. Geodin Acta 19(1):51–71. doi:10.3166/ga.19.51-71

Peulvast JP, Claudino Sales V, Bétard F, Gunnell Y (2008) Low post-Cenomanian denudation depths across the Brazilian Northeast: implications for long-term landscape evolution at a transform continental margin. Global Planet Change 62:39–60. doi:10.1016/j.gloplacha.2007.11.005

Projeto RadamBrasil (1981a) Levantamento integrado dos recursos naturais do Brasil. Folha Fortaleza. Ministério das Minas e Energia-MME, Brasilia

Projeto RadamBrasil (1981b) Levantamento integrado dos recursos naturais do Brasil. Folha Jaguaribe-Natal. Ministério das Minas e Energia-MME, Brasilia

Saadi A, Torquato JR (1992) Contribuição à neotectônica do Estado do Ceará. Rev Geologia 5: 5–38

Santos RV, Oliveira CG, Parente CV, Garcia MGM, Dantas EL (2013) Hydrothermal alteration related to a deep mantle source controlled by a Cambrian intracontinental strike-slip fault: Evidence for the Meruoca felsic intrusion associated with the Transbraziliano Lineament, Northeastern Brazil. J S Am Earth Sci 43:33–41

Sénant J, Popoff M (1991) Early Cretaceous extension in northeast Brazil related to the South Atlantic opening. Tectonophysics 198:35–46

Teixeira AL, Gaucher C, Paim PSG, Fonseca MD, Parente CV, Silva Filho WD, Almeida AD (2004) Bacias do estágio da transição da Plataforma Sul-Americana. In: Mantesso Neto V, Bartorelli A, Carneiro CDR, Brito-Neves BB (eds) Geologia do Continente Sul-Americano: Evolução da Obra de Fernando Flávio Marques de Almeida, Beca, pp 487–537

Thomas MF (1989a) The role of etch processes in landform development. I. Etching concepts and their applications. Z Geomorph NF 33(2):129–142

Thomas MF (1989b) The role of etch processes in landform development. II. Etching and the formation of relief. Z Geomorph NF 33(3):257–274

Vauchez A, Neves S, Caby R, Corsini M, Egydio-Silva M, Arthaud M, Amaro V (1995) The Borborema shear zone system, NE Brazil. J S Am Earth Sci 8(3–4):247–266

Chapter 4
Stepped Surfaces, Palaeolandforms and Morphostratigraphy: Reconstructing the Long-Term Landscape Evolution Since the Mesozoic

Abstract The present relief reflects a juxtaposition of highly contrasting elements of various ages. The landscape formed in the context of a fluctuating but continuously falling base level since the Cenomanian. After formation of the transform passive margin in Aptian times, landscape development was further driven by regional swell-like uplift. The seaward flank of this broad monocline was eroded, whereas the post-Cenomanian uplift caused an inversion of the Cretaceous basins and generated a landscape in which the most elevated landforms correspond either to resistant Mesozoic sedimentary caprock, or to eroded stumps of syn-rift Cretaceous footwall uplands. Some topographic surfaces at low elevations are Mesozoic land surfaces that became re-exposed in Cenozoic times. Denudation in the last 90 Myear never exceeded mean rates of 10 m Myear^{-1} and exhumed various Cretaceous stratigraphic unconformities. Obtained by morphostratigraphic methods, these results differ significantly from published AFT-derived estimates reported from Palaeozoic and Mesozoic sandstones of the Araripe basin, implying burial by a considerable thickness of younger section and, later, 1.5 km of post-rift denudation, i.e. two to three times as much as our maximum 0.6–0.7 km estimation. The likelihood of such a scenario is discussed, in the light of the exhumation of surrounding basement surfaces, which probably began very early, as shown to the northwest of the Chapada do Araripe by the presence of widespread laterites of probable Paleogene age. The distribution of these paleosoils indicates an early beginning of the basin topographic inversion, which might correspond to the first post-rift denudation episode reported by the AFTA. The second one, in the Oligocene and later, would correspond to the major stage of river incision and basin inversion. The origin of the discrepancy between the two sets of results is discussed, as well as the causes and exact timing of post-Cenomanian crustal upwarping.

Keywords Basin inversion · Denudation rates · Epeirogeny · Eustatism · Laterite · Palaeogeography · Palaeolandforms · Stepped surfaces · Soil cover · Uplift

J.-P. Peulvast and F. Bétard, *Landforms and Landscape Evolution of the Equatorial Margin of Northeast Brazil*, Springer Earth System Sciences,
DOI 10.1007/978-3-319-18203-2_4

4.1 Stepped Paleosurfaces and Related Soils as Indicators of Long-Term Landscape Evolution

4.1.1 The Identification of Stepped Surfaces

A series of topographic profiles transverse to the continental margin (Fig. 4.1) outlines the treads of two main topographic levels: a low plain between 0 and 300 m a.s.l., sloping gently seaward (average slope angle <0.1°), and the discontinuous elements of a high plain between 750 and 1100 m a.s.l. The latter lacks any well-defined slope, even in the Central Ceará Highlands where the high topography occurs near the coast (Fig. 4.1, profiles 8–10). Profiles east of the Senador Pompeu Shear Zone (SPSZ) (Fig. 4.1, profiles 1–4 and 5–7) highlight a deep inward extension of the embayed lower plain ('Sertaneja' surface) as far as the foot of the Borborema and Araripe plateaus, with residual bedrock landforms standing on the plain. To the west, except in Central Ceará Highlands (Fig. 4.1, profiles 8–10 and 11–13), the low plain also exists with a similar seaward gradient (~0.1°). The Ibiapaba plateau is characterized by the regular westward dip slope of its sedimentary strata but forms an elevated flat surface at ~1000 m a.s.l. This is similar to the other major summits of the western compartment and suggests that the plateau was once continuous between the Ibiapaba scarp and the outliers.

These flat levels are often separated from each other by well-defined escarpments or slopes, but surfaces of intermediate altitudes are also identified, standing as low plateaus (Chapada do Apodi; 80–150 m), or forming rolling and dissected pedestals to ridges and crests of the Central Ceará and Borborema highlands

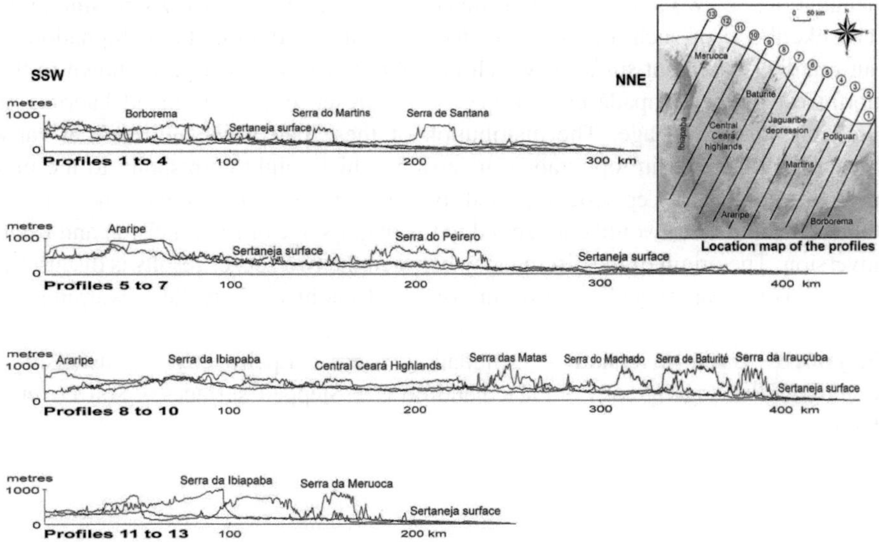

Fig. 4.1 Topographic profiles normal to the continental margin. *Source* SRTM DEM

(600–700 m). On the basis of our morphostructural study (Chap. 3), the two main storeys appear as composite features. The high level is composed of uplifted and dissected surfaces whose various origins are reflected by slightly different altitudes and forms. Below, the Sertaneja surface is made of interior depressions and corridors all connecting seaward in a wide and low piedmont surface partly buried by Late Cenozoic sediments and then slightly dissected.

4.1.1.1 Buried and Exhumed Erosion Surfaces

Planation surfaces (peneplains, pediplains or etchplains) form valuable set of landmarks for reconstructing landscape evolution and analysing tectonic deformations, but the problem most often encountered concerns their age (Watchman and Twidale 2002; Peulvast et al. 2009). This age may be found if a surface is preserved within a stratigraphy, by dating the rocks immediately above and below, but in most cases, exposed surfaces prevail and only a maximum age is yielded by the bedrock upon which they developed. Several planation surfaces form valuable landmarks in the study area, in spite of their dissection or degradation, or of the lack of dated sedimentary covers. Identification of synchronous remnants throughout a vast area is the most favourable case for evaluating tectonic deformations over a given period. Identification of diachronic surfaces or stepped sets of surfaces also gives clues on vertical movements and cyclic or non-cyclic types of geomorphic evolution.

The *infra-Paleozoic (post-Brasiliano) planation surface* is the oldest of these landmarks. It is identified below the Serra da Ibiapaba, in the central and eastern Araripe basin, and around scattered outliers, mainly in west and south Ceará. This surface, an equivalent of the oldest planation surface recognized in western Africa around the Taoudeni and Voltaian basins (Michel 1973; Thomas 1980), forms a major unconformable contact between the basement and the post-Brasiliana cover. In western Ceará, it is uplifted to more than 600 m only 50 km from the coast and records the same westward tilting as the overlying sandstone cover. Together with this cover, it is locally downfaulted along the Transbrasiliano shear zone (Santana do Acaraú graben: see Sect. 3.1.2).

Downwarping is stronger below the central and eastern Araripe basin; here, the total deformation of the surface, between +600 and −2000 m or lower in the deepest grabens records the effects of the Neocomian rifting, but also those of the post-rift subsidence, minus the amount of later uplift (see Fig. 3.5b, UV). The role of this surface in the present landscape is minor, since rather than being preserved after exhumation from its cover, it is generally intersected by younger erosion surfaces. East of the Araripe basin, locally strong dips might explain poor conditions of exhumation. The occurrence of wider remnants of this surface may be suspected in less deformed regions, e.g. NW of the Araripe basin, where basement plateaus partly capped by a much younger lateritic cover lie in the approximate continuation of the buried surface between 600 and 700 m a.s.l.

Although no Paleozoic sediments were found in rift structures outside the Araripe basin (Matos 1992, 2000), the existence of other remnants of an old surface is also suspected outside the basins. South of the Potiguar basin, the hanging pedestal of the sandstone buttes of Portalegre, Martins, João do Vale and Santana, and the Pereiro plateau might belong to such a surface, if we assume an old, possibly pre-rift, age for these sandstones (Peulvast et al. 2008). Preserved on a few kilometres wide below sandstone and laterite ledges and above the low surrounding pediplains, its rolling etch topography nearly coincides with the top of the basement exhumed from the thin overlying sandstone layer. Partly dissected or degraded, this surface displays numerous etch basins and fracture valleys, a few tens of meters deep, and rocky domes or bornhardts, carved into irregularly weathered and strongly kaolinized granites. In places, more resistant gneissic or granitic facies remain fresh looking up to the bevelled contact. Marble lenses form residual peaks in front of the scarp above the Apodi-Lajes depression. Such similarities with the infra-Paleozoic surface might suggest possible origin or age connections. They were not found around and below Cretaceous sediments (for instance below those of the Chapada do Araripe: Peulvast and Bétard 2015).

Similar features exist on other high basement surfaces (Pereiro plateau), often in the absence of any sedimentary outlier (Baturité and Meruoca plateaus). Their interpretation as remnants of the post-Brasiliano planation surface—suggested on the Meruoca plateau by the graphic reconstruction of this surface (see Fig. 3.13)—remains speculative.

The identification of elements of *Cretaceous planation surfaces* at various topographical levels is one of the most original results of our study. Bore-holes and geophysical studies have shown that rift sediments of the margin (Matos 2000) and of the Cariri-Potiguar rift zone outside the Araripe basin (Matos 1992) rest on the basement. Therefore, the pre-Cretaceous (or pre-rift) topography had preserved post-orogenic, pre-rift sediments only in west and northwest Ceará (Parnaiba basin and vicinity), in south Ceará (Araripe basin), and possibly in smaller areas between the Potiguar and Rio do Peixe basins (Serra do Martins). Very few fragments of this surface are presently found outside the rift basins, owing to strong uplift and tilting of their shoulders.

Analysing the geometry of unconformities yields information on erosion surfaces shaped during rifting, oceanic opening, and later stages of the margin formation. Sediments of the transitional and post-rift stages generally rest in unconformity on buried and bevelled rift structures and onlap the basement that forms the rift shoulders (Matos 1992; Martill 1993; Hegarty et al. 2002). To the south, the same is observed on the Recôncavo-Tucano-Jatobá rift, where the alluvial conglomerates and sandstones of the Marizal Formation (Aptian) lie upon an undulating eroded surface underlain by tilted rift sediments and formed between 124 and 120 Myear (Magnavita et al. 1994).

In the Araripe basin, a 20 Myear gap and a regional unconformity separate the transitional series of Aptian age from the rift series and structures, reflecting uplift and widespread erosion that had taken place soon after rifting or during its last stages, before ~114 Myear (Hegarty et al. 2002; Peulvast and Bétard 2015). Later subsidence of a large EW basin, possibly connected with the Potiguar basin (Arai 1999), allowed sedimentation beyond the rift limits. Among the 380–400 m of post-rift sediments whose remnants form the Chapada do Araripe, the lower lacustrine sediments of Aptian age unconformably lie on the older series, and bury incompletely bevelled basement highs (Martill 1993). They are covered by evaporites (gypsum), and fossiliferous deltaic silts and sands (Santana Formation). Evidence of increasing salinity within this formation suggests that marine water could enter the basin on occasions (Baudin and Berthou 1996; Neumann 1999), possibly from the Parnaiba basin (Andrade et al. 2002). A low altitude landscape of lakes or lagoons surrounded by faint hills persisted during the Albian before being finally covered in the Cenomanian by the Exu fluvial conglomerates and sandstones, deposited from the SSE (Hegarty et al. 2002), and onlapping the basement to the W and NW.

This last unconformity surface is uplifted to 600–700 m and partly exhumed south of Campos Sales, where it intersects at low angle the exhumed post-Brasiliano surface (see Fig. 3.5b, UV). Therefore, we consider the Campos Sales plateau as a polygenetic surface, well preserved because of its location at the divide between the Jaguaribe and Parnaiba watersheds, far from the base level. To the north, along the Serra da Ibiapaba, the infra-Paleozoic surface was destroyed as fast as it was exhumed. Parts of it were twice buried and then exhumed or slightly reshaped, before being capped, in post-Cenomanian times, by a lateritic cover developed after dismantling of the surrounding plateaus, and then degraded by downcutting and head retreat of tributaries of the Jaguaribe river. In this area, the most achieved erosion surface is the Pre-Cenomanian (Albian) one, possibly perfected in the plane of the last Cenomanian filling surface at the periphery of the basin (Fig. 4.2a).

4.1.1.2 Palaeo-Piedmont Surfaces

Parts of the Cretaceous and later erosion surfaces are piedmont surfaces shaped into rift shoulders and in the onshore passive margin, in front of retreating or dismantling fault or flexure scarps.

In the Potiguar basin, widespread erosion of the rift shoulders and syn-rift sediments had taken place before the Aptian, followed by thermal subsidence and deposition of sediments on large areas, beyond the limits of rift deposition. In the onshore basin, the 400–800 m thick Açu sandstones (Albian-Cenomanian) rest unconformably on rift or transitional deposits and bevelled basement highs (Castro et al. 1981; Neves 1989; Araripe and Feijó 1994), extending up to 20–30 km beyond the master faults (Apodi, Carnaubais and Portalegre-Gangorra faults; Fig. 4.3).

South of the Chapada do Apodi, the 30–60 km wide erosion surface that separates the limestone cuesta from the dissected scarp (Portalegre, Martins, João do Vale and Santana buttes or mesas) is identified as deriving from the exhumed pre-Cenomanian surface, still bearing a few outliers of Açu sanstones, west and southwest of Apodi (Peulvast and Claudino Sales 2004; Fig. 4.3). Its development, linked with scarp retreat on ca 20–30 km from the Carnaúbais, Apodi and Gangorra faults, occurred on a period of 20–30 Myear, during rifting and transitional stages,

(a)
Cenomanian Surface
Albian Surface
"Sertaneja" Surface
(b)
Exhumed, granitic paleo-inselberg
Cenomanian sedimentary cover (sandstone of Açu Formation)
(c)
Barreiras accumulation surface
Oligocene volcanic neck (Serrote Caruru)

◀ **Fig. 4.2** Paleosurfaces and paleolandforms in the northern Brazilian "nordeste". **a** South rim of the Chapada do Araripe, east of Exu, looking west: the Cenomanian Exu sandstone overlooks a narrow and dissected granite step above the Sertaneja depression, eroded into the uplifted basement (*foreground*). This picture shows three of the surfaces described in the text, grouped in two main topographic levels. The Cenomanian structural surface, and the Albian surface on which the thin sandstone layer was deposited, belong to the highest level, whereas the Sertaneja surface is part of the lowest. **b** Quixéré, lower Jaguaribe valley: granitic paleo-inselbergs partly exhumed from the Cretaceous sedimentary cover in front of the cuesta of the Chapada do Apodi. **c** Serrote Caruru, a phonolitic neck (~35 Myear) surrounded by argillaceous sands of the Barreiras Group and protruding through the tabuleiros on the west side of the Pacoti river. In the distance (SW): the Serra da Aratanha. *Photographs* J.P. Peulvast

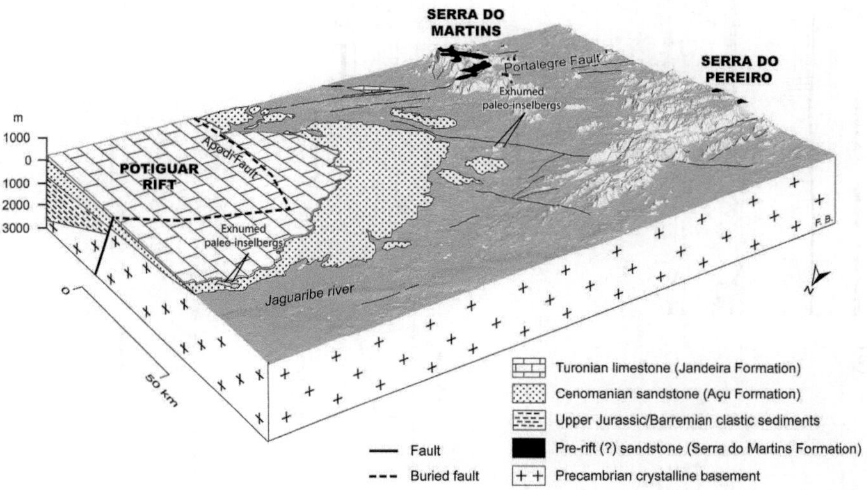

Fig. 4.3 Geological block-diagram of the southern shoulder of the Potiguar rift. The erosion surface that separates the Chapada do Apodi from the dissected scarp (Martins, Pereiro) is interpreted as a paleo-piedmont (i.e. the exhumed pre-Cenomanian paleosurface and its paleo-inselbergs), still bearing a few outliers of Açu sanstones at short distances from the erosional scarp. 3D-relief from the SRTM DEM and geology from CPRM (2001)

before the onset of the Açu sandstone deposition. This observation implies denudation by localised downwearing of at least 0.5–1 km of rock or more within the basement (Fig. 4.4), and/or scarp retreat at the rate of ca 1 km $Myear^{-1}$.

A similar exhumation pattern is observed along the subsequent low Jaguaribe valley, where granitic and quartzitic inselbergs are exhumed or still partly buried respectively in the front and on the backslope of the retreating cuesta (Fig. 4.2b: photo). Here, the exhumed pre-Cenomanian surface, reconstructed on geological cross sections ("Jaguaribe surface"), almost coincides with the Sertaneja surface west of the Jaguaribe River, at least over 40 km in width (Fig. 3.5b, OP, QR).

The low planation surface partly buried below the Cenozoic Barreiras sediments (the "pre-Barreiras surface") can also be considered as a palaeo-piedmont surface, whose inner parts are exposed at low angle with the plane of the coastal sedimentary

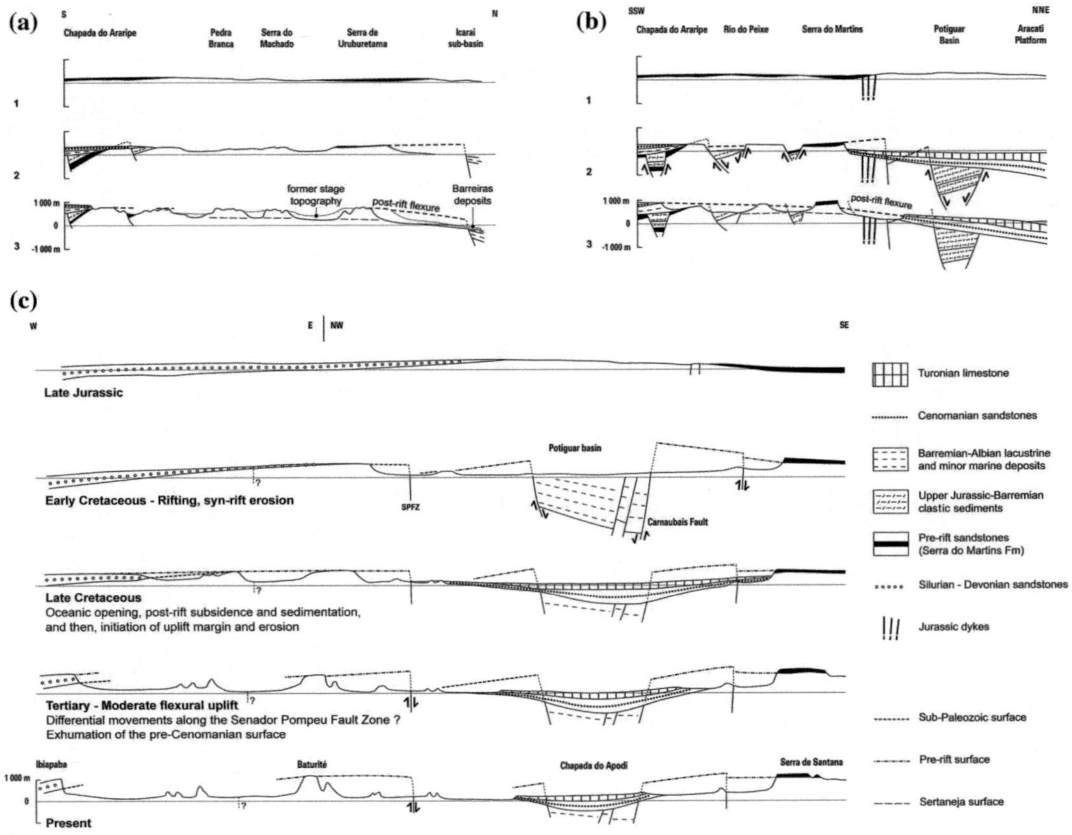

Fig. 4.4 Reconstruction of the morphotectonic evolution of the northern Brazilian "nordeste" since pre-rift times. Synthetic transverse profiles of the Equatorial margin on the west (**a**) and east (**b**) sides of the SPFZ, and through the Potiguar rift and its shoulders (**c**). From Peulvast et al. (2008)

plateau. Forming the pedestal of the marginal scarp and of numerous inselbergs, it is well identified around the contacts between the sediments and their substrate, near the coast and along valleys incised down to the basement. Generally flat, it is locally more irregular, with low rounded hills (north of the Serra de Baturité), isolated quartzitic hills, even at the coast (Jericoacoara) and quartzite or limestone inselbergs protruding through the sedimentary cover (Cascavel). Around Fortaleza, the Barreiras sediments surround volcanic necks towering above the buried surface (Serrote Caruru: Fig. 4.2c), and locally fill depressions of the underlying basement, which may be deeply weathered, suggesting incomplete pre-Barreiras stripping of older saprolites (Cascavel area: Cunha and Castelo Branco 2002).

In western and central Ceará, this surface is geometrically the distal continuation of the Sertaneja depression. It clearly post-dates Cretaceous faulting along the Transbrasiliano shear zone. In eastern Ceará, its profile roughly coincides with the geometric continuation of the Jaguaribe surface, but at regional scale, along the coast, it bevels both the basement and the Late Cretaceous sediments of the Chapada do Apodi, therefore appearing as a post-Campanian surface. The deposition of Barreiras argilaceous sands around volcanic necks of Oligocene age in the Fortaleza area (Almeida and Carneiro 1998; Fig. 4.2c) implies that at least parts of the underlying surface were shaped or re-shaped after erosive destruction of the former volcanoes, formed 35 Myear ago.

The burying of bevelled and dissected topographies carved into highly deformed post-Campanian lateritized sediments on the coast of the Potiguar basin near Icapui (Fig. 4.5) suggests that parts of this surface record a complex erosional history, whose last stages have been controlled by neotectonics (Bezerra et al. 2001; Sousa et al. 2008a) and Late-Cenozoic sea-level falls (Shimabukuro and Arai 2001; see Chap. 5). Therefore, far from being simple, this piedmont surface is a composite and diachronic one, whose buried as well as exposed elements intersect each other at low angles and locally comprise dissected surfaces (Cascavel area).

4.1.2 Soils, Weathering Crusts and Superficial Deposits: Chronological Consequences

Most surfaces lack equivalents of the well dated superficial deposits or weathering crusts that help analysing stepped landforms in the Guianas (Krook 1975) or in western Africa (Mali, Guinea: Michel 1973; Gunnell 2003). However, remnants of lateritic crusts are recognized in a few areas, on small mesas capped by the Serra do Martins sanstones, on the Serra do Pereiro (Peulvast and Claudino Sales 2004; Bétard 2007; Gurgel et al. 2013), on the Chapada do Apodi (Ponta Grossa-Peroba cliffs: Fig. 4.5; north of João Camara, RN), and on palaeosurfaces of the Campos Sales-Araripe area (Bétard et al. 2005). The first occurrence corresponds to remnants of a high structural surface (650–700 m), the others are recognized at lower altitude on various tabular or bevelled structures. Some interpretations propose age and origin similarities between the Serra do Martins Formation, the post-Campanian

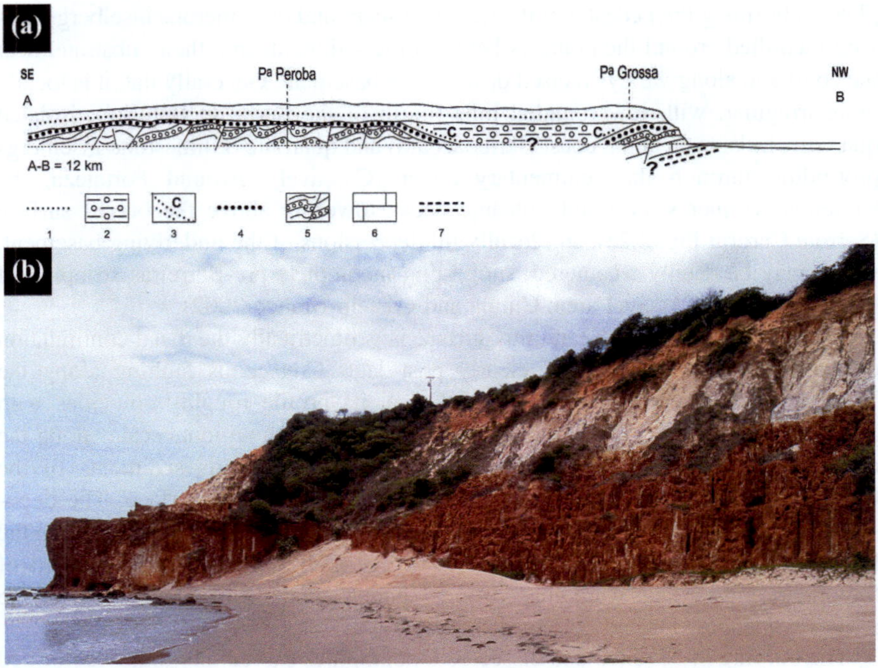

Fig. 4.5 Complex burial-erosion history revealed in sea cliff outcrops between Icapui and Ponta Grossa, east Ceará. **a** Sketch profile, out of scale (approximate height: 20–30 m). *1* Palaeodunes; *2* horizontal layers of reddish argillaceous sand (Barreiras sediments s.s. *question mark* possible fluvial origin); *3* slope deposits on the flanks of the Redonda buried depression (palaeovalley *question mark*); *4* allochtonous laterite; *5* strongly deformed (faulted and folded) and lateritized sandstones and conglomerates; *6* Jandaira limestone (Turonian-Campanian); *7* Açu sandstone, Cenomanian (*question mark*). **b** The Redonda buried depression or palaeovalley and its east slope (between Ponta Peroba—visible on the *left side*—and Ponta Grossa). Below a thick unconformable palaeodune material, contact between horizontal layers of reddish argillaceous sand (*foreground, right*; Barreiras sediments s.s. *question mark* possible fluvial origin) and strongly deformed (faulted and folded) and lateritized sandstones and conglomerates of uncertain age (at the *back*). *Photographs* J.P. Peulvast

sediments of Peroba-Ponta Grossa, and the Barreiras Formation (Sousa et al. 1999). In fact, similarities between weathering formations do not imply uniform ages for underlying sediments and surfaces, and morphostratigraphic relationships do not favour this interpretation.

So far, ferricretes have not been systematically mapped in the study area. Only the long durations (millions of years) generally involved in laterite formation (Lappartient 1985; Tardy and Roquin 1998; Widdowson and Gunnell 1999; Lageat and Gunnell 2001) suggest pre-Quaternary ages for them. According to available palaeogeographical maps (Tardy and Roquin 1998), conditions favourable to laterite formation might have occurred in inner Northeast Brazil only during Early Tertiary (Eocene), whereas later and drier periods (Neogene: Thiry et al. 1999) would have only allowed sporadic formation of ferruginous gravels in lixisols at low altitude. In more humid coastal areas, lateritic soil formation has been recognized in the Barreiras sediments for the mid-Miocene to late Quaternary period (Rossetti et al. 2013).

In most other regions, rather thin soils and saprolites prevail, appropriate to the present semi-arid climatic conditions (leptosols, luvisols and cambisols), except on sand or sandstone surfaces of high or low altitudes (lixisols and arenosols: Bétard 2007) and in the more humid coastal strip and mountain slopes, where thicker tropical soils (lixisols, acrisols) and red or yellow ferralsols occur. Most of these soils reflect the present distribution of bioclimatic conditions and even those which cover erosion or structural platforms cannot yield valuable age information on the large scale morphology, often inherited from much older times. However, the regional distribution of soils is highly reflective of the stepped pattern of landforms and associated denudation rates (Bétard and Peulvast 2013). Whereas only non-lateritic soils developed from shallow, grus-type weathering mantles occur on the Sertaneja surface marked by semi-arid conditions since the Miocene, relict laterites inherited from older (Paleogene) paleoclimates remain currently scarce in the landscape and are confined to ancient summit surfaces and/or in positions of major drainage divides (Serra do Pereiro, Araripe-Campos Sales area: Bétard and Peulvast 2013). This organization of soil covers is symptomatic of contrasted denudation rates, with higher rates on the low-altitude Sertaneja surface steering the landscape towards the stripping of most of the paleoweathering mantle, and favouring the development of a primary (non-lateritic) fersiallitic pedogenesis under the newly established semi-arid conditions, in existence at least since the last 13 Myear (Harris and Mix 2002; Bétard et al. 2009; Bétard 2012).

Remnants of waste deposits identified on parts of these surfaces—alluvial fans, terraces, pediment covers—are more readily interpretable. According to the geological map of Ceará (CPRM 2003), such deposits of argillaceous, often ferruginous, sands and gravels of fluvial to colluvial origins are found either on high basement surfaces (southwest Ceará), or on slightly dissected pediplains at low altitudes. Though the authors of the geological map include Holocene deposits among these sediments, the most of them probably represent older formations, even on low surfaces, since a slight entrenchment of valleys now cuts them from their source areas, scarps and inselberg slopes located up to 20 or 30 km away (Baturité and Quixadá areas: Bétard 2007; Peulvast and Bétard 2013).

4.1.3 Interpretation of the Stepped System of Landforms

Our chronology of the stepped surfaces is far from matching those of classical models (see Sect. 1.3.2). However, as in these cyclic models, the *high plateaus*, between 700 and 1100 m, generally correspond to the oldest surfaces (Peulvast and Claudino Sales 2004). Mainly located at the periphery of the Jaguaribe-Piranhas amphitheatre, they are also found in more central parts, forming narrow mesas, buttes or flat-topped mountains, some of which bear the highest peaks.

These residual plateaus (Fig. 4.6) include (i) dissected remnants of structural surfaces of unknown age, in which the topography coincides with the upper surface of a resistant sedimentary layer (Ibiapaba, Serra do Martins); (ii) a structural surface of Cenomanian sandstone corresponding to the weakly degraded top of the post-rift Araripe series; (iii) exhumed patches of the sub-Paleozoic (post-Brasiliana) erosion surface, best exposed in southwest Ceará where the Ibiapaba Paleozoic cover rocks have been stripped back by erosion; and (iv) exhumed tracts on the northwest margin of the Araripe basin of a sub-Albian erosion surface, capped by laterite. The plateaus that collectively form the high plain carry the most deeply and intensely weathered materials of the region, regardless of parent rock. For instance, exhumed kaolinitic weathering fronts are preserved on the southern footwall of the Potiguar basin (Serra do Martins-Serra de Santana). West of Fortaleza, elevated remnants of deeply kaolinized weathering profiles also cap massifs closer to the coast such as Meruoca and Baturité (900–1000 m a.s.l.).

Therefore no simple and synchronous high surface is identified that might correspond to Ollier and Pain's "palaeoplain" (1997). We rather find remnants of old surfaces of varied origins, possibly of a pre-rift polygenetic and differentiated surface, unequally uplifted. Its preservation on rift shoulders depends on the presence of resistant rocks or of resistant sedimentary covers and ferricretes, still present or not.

The low plain (Sertaneja surface) is also a mosaic of erosion surfaces of different ages, some of them traceable to exhumed stratigraphic unconformities (e.g. the exhumed sub-Cenomanian or sub-Albian surface of the Aracati–Potiguar area, well preserved around outliers of Açu sandstone southwest of the Potiguar basin). This exhumed surface is also exposed around the lower Jaguaribe valley, with partly exhumed granitic inselbergs (Fig. 4.2b). As in the Recôncavo–Tucano–Jatobá rift (Magnavita et al. 1994) and in the Araripe basin (Peulvast and Bétard 2015), where this exhumed sub-Cenomanian erosional plain is also reported, its extensive development testifies to intense erosion during the opening of the Potiguar failed rift. Contrary to previous landscape development scenarios for this region (see Chap. 1), this also implies that the lower-lying topographic surfaces of the northeast Brazilian topographic staircase are not systematically the youngest (Fig. 4.6).

The lateral topographic continuity between this exhumed surface and the low erosion surfaces that surround the mountains of northern Ceará (Fig. 4.4) might also suggest Cretaceous ages for these surfaces, but no Cretaceous deposit is known here. Located on the northwest side of the Senador Pompeu fault zone and belonging to the uplifted shoulder of the Cariri-Potiguar rift zone as well as to that of the Ceará transform basin and margin, this area remained outside the Aracati-Potiguar subsidence area. The presence of a probable Cretaceous, syn- or post-rift planation surface at high altitudes on remote parts of the rift shoulder suggests that at least part of this area underwent significant post-rift uplift and that the low Sertaneja pediplain developed here by cyclic entrenchment from the early bevelled coastal strip. Such a difference with the subsiding Jaguaribe area implies differential post-rift movements along the Senador Pompeu fault zone and other related faults (Figs. 4.4 and 4.6). The wide development of the low surface on both sides of this

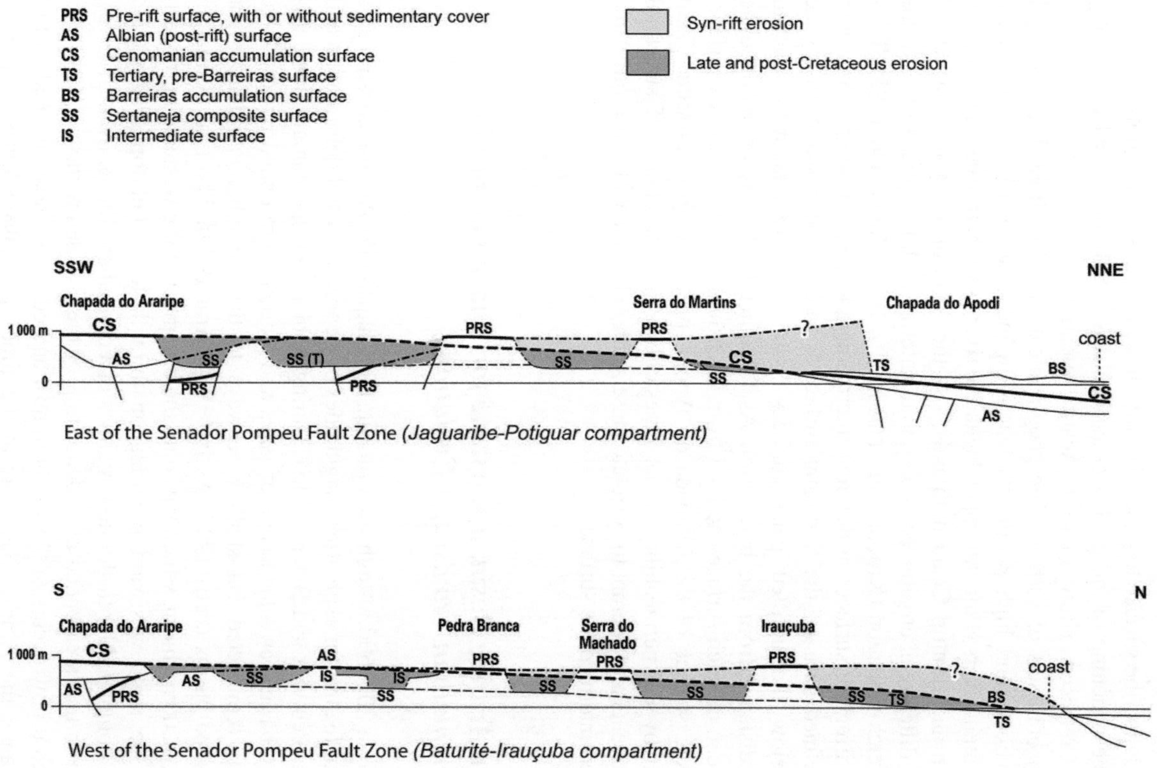

Fig. 4.6 Synthetic transverse profiles of the Equatorial margin of Northeast Brazil showing the tiered landform systems on both sides of the Senador Pompeu fault zone. From Peulvast et al. (2008)

fault, around the Baturité massif and Quixadá may be explained by slow and diffuse movements in this former shear zone, allowing the ongoing process of planation of the nearby Jaguaribe surface to proceed in this area in spite of its uplifting trend. To the west, the widening of the piedmont surface was locally facilitated by faulting during the Cretaceous rifting or later (Acarau re-entrant).

Between both sets of surfaces, the geographical and altitudinal distribution of depression floors and high pediments that form the pedestal of residual ridges, around the main rivers (Poti and upper Jaguaribe rivers and their tributaries), mainly in the south and Central Ceará highlands, is rather controlled by local structural factors and differential erosion, according to a pattern which is typical of landforms shaped by etch-planation (Kroonenberg and Melitz 1983; Zonneveld 1985). They form more irregular surfaces at different intermediate levels, whose only link with the well-defined parts of the Sertaneja surface is to have been shaped in cyclic regimes below the uplifted palaeosurfaces, with lesser entrenchment due to increasing distance from the base level. As for other landforms of intermediate altitudes—e.g. the etch surfaces derived from the exhumed pre-Cretaceous surface, or the karstic surface of the Chapada do Apodi, preserved from dissection and degradation by the permeability of its limestones—the importance of structural controls prevents any attempt to correlate these scattered elements to each other in a well-defined intermediate surface.

4.1.4 The Araripe Basin, a Keystone for Understanding Post-cenomanian Landscape Evolution

Bounded by a 200–600-m-high escarpment, the Chapada do Araripe constitutes a key regional feature for estimating magnitudes of Cenozoic crustal deformation and denudation (Peulvast and Bétard 2015). During Albian to Cenomanian times, this sedimentary basin was a landscape of lakes and lagoons surrounded by low hills, intermittently connected via shallow seaways to the Parnaiba, Potiguar and/or Tucano–Jatoba basins (Petri 1987; Assine 1994; Arai 1999). The Exu Formation (90 Myear, <200 m thick), which laps onto the basement to the west and northwest, represents the rapid westward progradation through low-relief topography of a braid-plain dominated by flashy flow regimes in a dry palaeoclimate (Martill 1993). Subsidence in that area ended after deposition of the Exu sediments.

Although it now occurs in the southernmost and currently most elevated part of the study area, the Araripe basin lay at the time at a palaeoelevation close to palaeosea level. Studies of the organic matter contained in the Albian sediments suggest that no significant overburden was ever removed by erosion from the exposed upper surface of the Exu caprock (Baudin and Berthou 1996; Arai 2000). This region has thus formed an almost uneroded topographic surface for the last 90 Myear. It logically follows that the Sertaneja plain developed in post-Cenomanian times by erosion of the seaward flank of a broad crustal upwarp that effectively inverted the Araripe and other Cretaceous basins, with drainage making inroads into the hinterland through older Cretaceous half-grabens and Brasiliano shear zones (see hypothetical profile of Cenomanian surface on Fig. 4.4).

4.2 Palaeogeography and Chronology of Long-Term Landscape Development

4.2.1 The Pre-rift Period (Jurassic and Older)

The pre-rift sedimentary cover is preserved in western Ceará and in the Araripe basin (Matos 2000; Fig. 4.7a). Palaeozoic sandstones, fluvial sediments of Jurassic age, also present in the Araripe and Recôncavo–Tucano–Jatobá basins (Matos 1992; Magnavita et al. 1994), are related to an early phase of crustal extension, local uplift and erosion recorded by the AFT history of Silurian and Ordovician samples of the Araripe region (Morais Neto et al. 2005–2006). Whether in the Potiguar rift zone or offshore, remains of this cover are absent north of the Patos shear zone. This suggests that, contrary to the Parnaiba basin in the west, the basement and its Palaeozoic cover have been eroded. In the hypothesis of a pre-rift origin of the fluvial sediments known as the Serra do Martins Formation, these layers which currently cap the highest topography south of the Potiguar basin, would be the exception. Their preservation would have been made possible by the formation of an E–W crustal sag or topographic depression south of the future Potiguar rift zone. Such a depression might have been linked to the magmatic activity (E–W dyke swarms) and the related crustal deformation that occurred between 145 and 130 Myear (Carneiro et al. 1989; Oliveira and Gomes 1996; Archanjo et al. 2000).

4.2.2 Intracontinental Rifting (Early Cretaceous)

The intracontinental and offshore basins of the study area are Neocomian–Barremian and Aptian, respectively (Fig. 4.7b). Whereas the Cariri–Potiguar basin already had become a failed rift by the end of Barremian times (Matos 2000), major deformation became subsequently located on the equatorial branch of the rift system, with the onset of E–W extension generating transtensional conditions in the Potiguar basin. In early Aptian times, this new tectonic regime formed NW–SE trending *en-échelon* synclines cross-cut by a diffuse fault pattern, without typical rift structures or sediments (Matos 2000). To the west, fluvial and deltaic sediments (Mundaú Formation, Lower to Middle Aptian, 1800–4000 m thick) were deposited in basins formed before oceanic opening (Condé et al. 2007). The intracontinental basins had also been filled by thick series—up to 4 km in the Potiguar rift—of early Neocomian to Aptian fluvial, deltaic and lacustrine sediments, reflecting deep erosion of the uplifted basement. The regionally unconformable transitional and post-rift series show that widespread erosion of the rift shoulders but also of rift sediments had

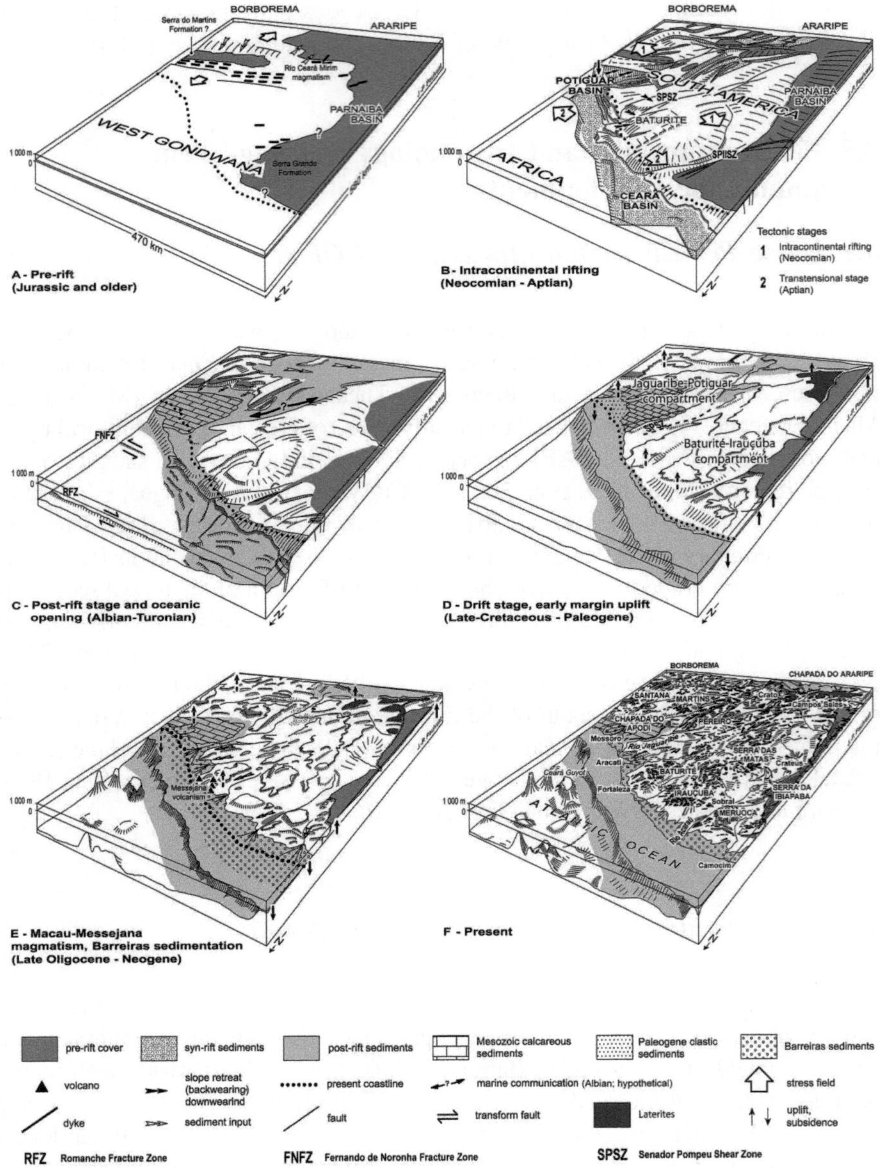

Fig. 4.7 Reconstruction of the long-term landscape evolution of the Equatorial margin of Northeast Brazil since pre-rift times. *Sources* Petri (1987), Matos (1992, 2000), Assine (1994), Peulvast and Claudino Sales (2003), and Pessoa Neto (2003). Note that topography in **f** corresponds to the current landscape. Landform outlines on diagrams **a–e** are inevitably a more sketchy depiction of the palaeolandscape. From Peulvast et al. (2008)

taken place soon after rifting, over a time span of ~20 Myear (Peulvast and Claudino Sales 2004). Similar palaeolandscapes developed in the Araripe basin (Martill 1993; Ponte and Ponte-Filho 1996; Peulvast and Bétard 2015).

4.2.3 Post-rift Stage and Oceanic Opening (Middle to Late Cretaceous)

By the time transtensional conditions had become dominant in the Equatorial Atlantic domain, the aborted Cariri–Potiguar rift zone was already undergoing thermal subsidence (Mello 1989; Ponte and Ponte-Filho 1996). Until the late Cretaceous, the rift zone became an area of widespread sedimentation, which was initially lacustrine or lagoonal but finally fluvial in the Araripe basin (Exu Fm, Cenomanian), and continental (Açu Fm) but later marine (Jandaira Fm, Campanian, ~80 Myear) in the Potiguar basin (Fig. 4.7c). Subsidence ended after the deposition of the Exu sandstone in the Araripe basin but has continued until present time in the offshore Potiguar basin.

In the onshore Potiguar basin, the presence of thick (up to 400 m) bioclastic calcarenites and calcilutites deposited in lagoons or shallow open seas during Turonian to early Campanian times (Jandaira Fm) shows that after a transitional stage characterized by estuarine or deltaic conditions, the terrigenous input decreased and erosion on basement topography became limited to chemical weathering and removal of solutes. Similar trends are known in the (offshore) Ceará basin (Costa et al. 1990; Beltrami et al. 1994). Synsedimentary dextral transcurrent movements deformed the Albian to early Campanian transgressive series of carbonates and shales, with a development of transpressional structures (folds, thrust faults, flower structures), but no coeval denudation was recorded onshore. The high sea levels that prevailed during this period (Haq et al. 1987; Arai 2000; Miller et al. 2005) were decisive in keeping the continental relief to a minimum. Moreover, the regional climatic conditions prevailing during the late Cretaceous were warm (Huber et al. 2002) and humid, favouring deep kaolinitic weathering (Schmitt 1999) and leaching of carbonates. Because of both syn- and post-rift erosion south and west of the Potiguar basin (Petri 1987) and of sediment aggradation in the basins, the heights of existing rift-flank escarpments were probably quite diminished by Campanian times. In summary, apart from a few upstanding residual massifs the landscape towards the end of the Cretaceous was predominantly flat and low-lying.

4.2.4 Regional Uplift, Erosion and Topographic Inversion (Late Cretaceous to Present)

The subsequent stages of uplift and erosion are key to understanding the present day scenery, but remain less well constrained by stratigraphy than the earlier evolution due to a ~50 Myear hiatus in onshore sedimentation between the Cenomanian and the Neogene. Post-Cenomanian basin inversion involved maximum surface uplift in the Chapada do Araripe area, with magnitudes steadily declining towards the coastline. The sedimentary record of denudation for that period is contained in offshore deposits and in the coastal belt of Barreiras sediments, suggesting a definitive shift of depocentres to the new Atlantic margin and a steadily buoyant and eroding hinterland (Fig. 4.7d).

Despite the neotectonic interpretations of the Cenozoic landscape proposed by Jardim de Sá et al. (1999) and Gurgel et al. (2013), neither post-rift faulting nor important local post-Campanian rock deformation have been detected outside the Potiguar basin. Exceptions exist along a few segments of the Carnaubais master fault of the Potiguar basin, and are involved in the formation of the most salient landforms of the Chapada do Apodi (Serra do Mel, Serra de Mossoró), as well as in spectacular faults and folds visible in sea cliffs but deprived of geomorphological expression on the plateau (Icapui-Ponta Grossa: Sousa et al. 2008a, b). Regional-scale flexural deformation was therefore the dominant style of crustal deformation outside the coastal strip. A sharp increase in clastic discharge peaking in late Miocene times (Pessoa Neto 1999, 2003) was interpreted by Morais Neto et al. (2000) as a consequence of uplift and erosion affecting the east Borborema province. Here we suggest that uplift affected a broader swell also extending to the west of the Borborema highlands, i.e. into Ceará. Dissection related to uplift and, finally, to late Cenozoic eustatic movements, formed the Chapada do Araripe and Chapada do Apodi by major and minor topographic inversion, respectively, and caused partial exhumation of the sub-Cenomanian palaeosurface (Fig. 4.7e).

In the Ceará basin, sediment influx from the continent seems to have remained moderate and often discontinuous. Only Cretaceous deposits are found on basement highs. The diversity of structural controls along the equatorial margin (Matos 2000), but also the strong asymmetry of drainage patterns observed in the Jaguaribe–Piranhas embayment (see Chap. 3), may explain the heterogeneity in depositional patterns. Whereas a non-depositional hiatus of 50 Myear separated the Cenomanian from the late Eocene sequences in the Piauí–Camocim sub-basin (Beltrami et al. 1994), a more continuous sediment flux is recorded in the Acaraú and Icaraí sub-basins, and at least six erosional or nondepositional events are recognized among Lower Paleocene to Lower Oligocene depositional units in the Mundaú sub-basin. The Oligocene recorded higher rates of sedimentation, at least in the Mundaú sub-basin (Cunha 1991; Condé et al. 2007).

In spite of relatively high sea levels prevailing until the Tortonian regressions (Haq et al. 1987; Miller et al. 2005), only terrigenous sediments were deposited on the coastal erosion surface. Whereas this Barreiras Formation, which is the onshore equivalent of the Miocene siliciclastic sediments of the inner shelf (Pessoa Neto 2003), is attributable to a response to crustal uplift, the relative increase in clastic supply from the hinterland is also probably linked to a marked shift towards aridity at that time, as suggested by Harris and Mix (2002) from a study of the ratio of oxide minerals in the terrigenous sedimentation of the Ceará Rise. We link the subsequent dissection of the Barreiras beds (Fig. 4.7e, f) to Pliocene and Quaternary sea-level fluctuations (see Sect. 5.4.4).

4.3 Uplift and Denudation as Driving Factors of Long-Term Landscape Development

4.3.1 Epeirogenic Deformation and Rates of Uplift and Denudation from Morphostratigraphic Data

Rates and amplitudes of tectonic movements were estimated on the southeast side of the SPSZ and in south Ceará from the altitudes of marine sediments, the most of which were deposited in shallow environments (Peulvast et al. 2008). In the Potiguar basin, the base of the Jandaira limestone (92 Myear) currently occurs between −500 m at the coastline near Tibau, and 60–120 m a.s.l. on the edge of the Chapada do Apodi. According to available curves, sea-level was at +140 or +80 m in early Turonian times (Haq et al. 1987; Miller et al. 2005), so deformation of this limestone layer records a post-Turonian subsidence of ~600 m at the coastline and 20 m at the landward periphery of the basin (or a slight uplift according to the curve by Miller et al. 2005). Based on the current elevation of marine Albian layers occurring up to 700 m above present sea level (Baudin and Berthou 1996; Neumann 1999), the Araripe basin in the remote hinterland was more vigorously uplifted. Assuming that sea-level rose from +150 to +220 m in Albian times (Haq et al. 1987), this suggests minimal post-Albian crustal uplift of 500–600 m, which is comparable to values proposed by Magnavita et al. (1994) in the Recôncavo–Tucano–Jatobá area. Because erosion has not significantly affected the Exu sandstone caprock (Baudin and Berthou 1996; Arai 1999; Peulvast and Bétard 2015; see also Sect. 5.3.2), the ~600 m of post-Albian crustal uplift in the area now forming the continental divide ca 300 km from the coast is also an estimate of long-term surface uplift of this part of the Brazilian shield.

Maximum post-rift denudation depths are provided by the maximum value of topographic inversion observed along the northern Serra da Ibiapaba and the eastern Chapada do Araripe, i.e. ~600 m near the city of Crato. The corresponding mean erosion rate (7–10 m/Myear) is similar to that of vertical movements.

Estimates of eroded mass onshore can only roughly be inferred from published data on offshore sedimentation (Fig. 4.8). Most of the sediments are trapped on the margin, which allows the system under investigation to be treated as relatively closed since Cenomanian times. An integration of post-Cenomanian denudation depths across the Jaguaribe–Piranhas embayment from the 600 m Araripe maximum to 0 m at the coastline would produce 1470 m of decompacted sediments distributed over the 120-km-wide offshore margin (shelf, continental slope, and continental rise). This calculation is based on mean densities of 2800 kg m^{-3} for the unweathered bedrock and 2000 kg m^{-3} for the terrigenous products. Sediment recompaction by a factor of 10 % (Boillot and Coulon 1998) obtains a 1320 m thick pile distributed over a mean accommodation width of 120 km. Given that the main drainage systems (Jaguaribe–Apodi–Piranhas: 73 % of the total study area) currently deliver their sediments to only 30 % of the linear extension of the margin (i.e., the offshore Potiguar basin), greater thicknesses of offshore sediment should

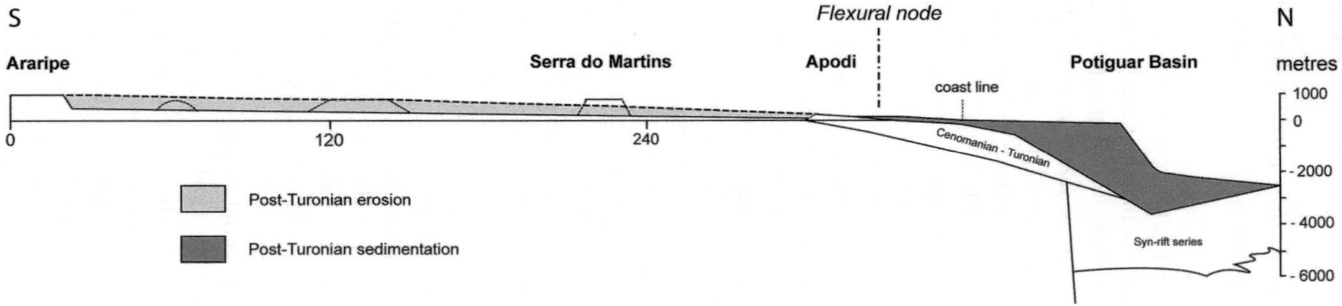

Fig. 4.8 Synthetic cross-section showing respective depths and thicknesses of post-Turonian erosion and sedimentation on the continental margin of eastern Ceará. Continental shelf and continental slope after Morais Neto (1999). Bathymetry from CPRM (2003). Hypothetical profile of continental rise from Boillot and Coulon (1998)

be expected in this segment. A rough estimate on that basis suggests that thicknesses of 3200 m for the post-rift series should exist in that depocentre, and up to 6000 m if deposited on the half width, mainly on the outer shelf and continental slope. These figures effectively correspond to known thicknesses of post-breakup series in that area (Morais Neto 1999; Matos 2000; Pessoa Neto 2003). Whereas more than 3000 m of post-breakup sediments were deposited in the offshore Potiguar basin (Morais Neto 1999), thicknesses only reach 1000–2500 m in the Piauí–Camocim sub-basin, and 1000–2000 m in the Mundaú sub-basin.

The mean thickness of the late Miocene and younger Barreiras and Tibau sediments Barreiras (Sediments) (150–200 m extending over a 100 km wide strip as measured on profiles of the Potiguar basin drawn by Pessoa Neto (2003) would represent a maximum eroded thickness of 50–70 m over a 300 km wide inland zone, at a rate of <10 m $Myear^{-1}$ over a $\sim$10 Myear interval. This matches the mean post-Cenomanian denudation rate estimated from the basin inversion criteria used in this study.

4.3.2 Denudation Depths and Long-Term Erosion Rates: A Discussion of Regional Data

Since Cenomanian times, the Ceará area of NE Brazil has shown evidence of shallow basin inversion. Uplift resulted in the topographic inversion of the post-rift basins, exhumation of buried surfaces/stratigraphic unconformities, dissection of the residual Cariri–Potiguar footwall uplands, and expansion of the erosional Sertaneja and coastal plains. Resulting long-term denudation rates in this setting were <10 m $Myear^{-1}$ and similar to values reported from the Appalachians (Gardner et al. 1993) or cratons in Africa (Bierman and Caffee 2001; van der Wateren and Dunai 2001; Gunnell 2003) and Australia (Stone and Vasconcelos 1999).

Confirmed by ^{10}Be cosmogenic analyses made on surface samples, soil profiles and river sediments of distinct geomorphological domains in the eastern Borborema province (Morais Neto et al. 2010a, b), the low post-rift denudation rates in NE Brazil are explained by a conjunction of four factors that also influence the conditions of scarp formation and evolution (Peulvast and Bétard 2013): (i) the low magnitude of crustal uplift estimated by the current elevation of marine Albian layers (Araripe, Apodi); (ii) the low amplitude and long wavelength of crustal deformation of an initially low-relief topographic surface, which promotes a phenomenon defined as 'morphological resistance' (Brunsden 1993a, b) in which the development of high angle slope systems favourable to intense erosion is impeded; (iii) the lithological heterogeneity of the basement and its cover, with resistant bedrock outcrops (e.g. sandstone, limestone, granite) explaining the widespread preservation of residual topography still upstanding on the Sertaneja erosional plain and occupying over $\sim$50 % of its surface area; (iv) the long-term semiarid climate in NE Brazil, probably in existence for the last 13 Myear at least (Harris and Mix 2002). According to the results of ^{10}Be cosmogenic analyses, very low erosion rates (0.1–13 m/Myear) are observed on sedimentary mesas and crystalline plateaus of the Borborema highlands, whereas they are higher (1–11 m/Myear) on lower surfaces, in particular the Sertaneja surface (Morais Neto et al. 2010a). Even only representing the last 1.5 Myear of geomorphic evolution, these data are in good agreement with our model of stepped surfaces preserving old features and thin sedimentary covers at the top of the plateaus (Serra do Martins Formation) without important changes, whereas faster erosion occurs at low levels, shaping the inner parts of the Sertaneja surface, close to the escarpments. They also confirm the importance of lithological controls and differential erosion in the development of this low surface and of the surrounding scarps and ridges (Morais Neto et al. 2010b).

Denudation rates were higher during the Cretaceous rifting and transitional stages, as indicated in the stratigraphic record by the rapid pre-Cenomanian erosion of large parts of the Potiguar footwall uplands, over a period of 20–30 Myear or less (Peulvast and Claudino Sales 2004).

New AFTA results published by Jelinek et al. (2014) in more southern parts of northeast Brazil, from the Borborema plateau to the Conquista and Jetiquinhonha plateaus through the Sertaneja Depression and the Chapada Diamantina (Bahia State), roughly match with the estimates we obtained north of the Borborema and in Ceará, although the rates are generally somewhat higher than ours. According to their data, correlated with offshore sedimentary record, an important phase of denudation coincides with rifting of the margin at 140–120 Myear in the Borborema region, with denudation rates up to 30–40 m $Myear^{-1}$. The estimated onshore erosion in the São Francisco region is 500–1000 m. Onshore erosion rates remained high during the transitional phase, with rates reaching 40–75 m $Myear^{-1}$, except in the Sertaneja Depression, where they were lower and where the apparent AFT ages are the oldest (180–350 Myear). In the post-rift phase I (110–40 Myear) defined in this work, 1.6 km of denudation would have been recorded in the Borborema plateau, with denudation rates of 10–30 m $Myear^{-1}$. Sedimentation rates are generally lower in the post-rift phase II (40–0 Myear), but denudation rates would be somewhat higher on the Borborema plateau (up to 40 m $Myear^{-1}$). The systematic differences in rates might reflect a different, more abrupt uplift style and a deeper dissection of the plateaus located on the eastern margin of northeast Brazil in comparison with our study area. In both cases, the distal parts of the low Sertaneja surface were already formed during rifting and the transitional phase, between residual rift shoulder stumps, and then weakly uplifted and eroded as the least uplifted parts of a broad monocline (Fig. 4.8).

4.3.3 Interpretation of Discrepancies Between AFTA and Morphostratigraphic Results in the Araripe Basin

Already considered as a good marker of basin inversion, the Chapada do Araripe constitutes a key regional feature for estimating magnitudes of Cenozoic crustal deformation and denudation and calibrating the results of thermochronological analyses. Here, the lack of any remnant of post-Exu sediments in the region, including in the post-rift series of the Jatobá-Tucano basin in the south (Magnavita et al. 1994), does not plead for the past existence of a younger cover, in spite of recent interpretations of thermochronological results according to which a slice of 1000 m or more would have existed over the present pile (Morais Neto et al. 2005–2006; Japsen et al. 2012), before being totally removed.

Studies of the organic matter contained in the Albian sediments suggest that no significant overburden was removed by erosion from the exposed upper surface of the Exu caprock (Baudin and Berthou 1996; Arai 2000). This point has been discussed in some interpretations of thermochronological studies based on analyses of fission tracks in apatites (AFTA), which suggest the past existence of a much thicker cover. The AFT-derived estimates reported by Morais Neto et al. (2005–2006) from Palaeozoic and Jurassic sandstones of the Araripe basin indicate palaeotemperatures of 70–85 °C during the Cenozoic, implying 1.5 km of post-rift denudation. This is two to three times as much as the maximum 0.6 km estimated in this study.

Indeed, a post-Cenomanian deposition would be in weak concordance with the interpretation given by Harman et al. (1998) and confirmed by Morais Neto et al. (2005–2006), of a thermal history indicating a cooling episode beginning between 100 and 90 Myear (late Albian-Turonian), i.e. towards the end of the fluvial sedimentation, and a later one in the Oligocene. The first cooling episode would correspond to an exhumation stage which seems hardly compatible with then ongoing sedimentation.

According to Morais Neto et al. (2005–2006), the VR (vitrinite reflectance in organic matter) value is reported as indicating maximum heating to 93 °C after the deposition of the Batateira layers (maximal burial of 2.2 km, but before 100 Myear, i.e. before the deposition of the Exu sandstone, which therefore does not seem to have been followed by further burying). The pre-Exu evolution must be taken into consideration in order to understand the signification of these data obtained on samples taken at various altitudes, in various layers. The Exu sandstone lies unconformably upon the older formations, including the basement, to the south and close to Crato (Ponte and Ponte-Filho 1996), and uplift and some denudation locally occurred before its deposition on the margins of the basin. Since the morphology (including exhumed and still buried paleolandforms) and structure of this basin are quite complex, with differential movements that persisted until the end of the post-rift period (Baudin and Berthou 1996), the interpretation of the thermal history of each sample is delicate and probably insufficient to reconstruct a regional history of sedimentation, uplift and erosion.

In the diagram of thermal histories presented by Morais Neto et al. (2005–2006), all samples show paleotemperatures of 40–60 °C before the later Cenozoic cooling event, interpreted as reflecting a denudation of 1.5 km of sediments. As shown by other recognized discrepancies between AFTA results and morphostratigraphic evidence (French Massif Central: Ricordel-Prognon et al. 2010; NW England, SW Scotland: Łuszczak et al. 2014) and discussed in various publications (Gunnell 2003; Peulvast et al. 2008, 2009; Wildman et al. 2014), these results remain in the uncertainty domain of the method. They may also reflect a change in the geothermal gradient in the Oligocene, which is the time of volcanism in surrounding regions

(Oliveira and Medeiros 2012). Late cooling stages from temperatures of ~60–70 °C, since 20–30 Myear, are also reported in the Borborema plateau and might reflect part of this denudation stage (Jelinek et al. 2014). However, these authors mention that such cooling events are well known to arise as artifacts of the numerical inversion of track-length data, although they assume an accelerated Neogene denudation of the margin. Such contradictory interpretations lead us to favor those given by Arai (2000) and Baudin and Berthou (1996), according to whom the sands and conglomerates of the Exu Formation which fill the depression mark the end of the sedimentary history in the Araripe basin.

Moreover, X-ray diffraction analysis of clay minerals in pyrobitumen shale samples taken in five gypsum mines in the Araripe basin (Neto et al. 2013) reveals expansive and interlayered clays with illite/smectite (I/S) values ranging from between 10–30 % of illite layers. Recent models indicate that "Smectite illitization occurs very early … at shallow burial depth…" (Lanson et al. 2009). The presence of I/S interlayered clays might also be interpreted as a result of intercalation of smectite with clastic clay minerals, which would also indicate shallow burying and long-lasting conditions of low temperatures.

The second caveat, which is also critical in the "episodic burial and exhumation model" developed by Japsen et al. (2012) for the post-rift evolution of the Brazilian Northeast, relates to the origin of the additional overburden inferred from AFT analysis. The upper Exu layers are horizontal on N–S profiles, reflecting the end of post-rift subsidence. An overburden of 1000 m or so would not have been deposited without further deformation of the underlying layers, unless spread over an area extending well beyond the present limits of the basin. The corresponding volume would have been huge, and its source remains highly uncertain. How would ~1000 m of additional sediment have been provided in a landscape shown by several authors (see Martill 1993) to have remained surrounded by low hills? Only the erosion of the Borborema highlands, to the east (the source area: Assine 1994), might have fed such a sediment mass in much larger quantities than the Exu sandstones. The other question is how would such sediments, of unknown nature (clastic or marine?), have spread over the backslope of an uplifting Great Escarpment (that is, if we consider the Borborema as part of such a structure) rather than on its oceanic side (where no such sedimentary record has been detected offshore). What could have been the topographic limits of the basin? And finally, to where would the waste generated by post-depositional stripping of the sediment mass have been conveyed during the later Cenozoic?

In our reconstruction of the regional geomorphic evolution based on the study of existing sedimentary vestiges and outliers (4.2), we indicate a discontinuous post-rift cover in large parts of Ceará and its surrounding regions, with temporary marine connections between the Parnaiba, Potiguar, Araripe and Recôncavo–Tucano–Jatobá basins in the Albian. Onshore, reconstructed geological sections indicate maximum thicknesses of up to 600–700 m close to the coast (Potiguar basin) and in the Araripe basin. These values are compatible with the volumes of material that the erosion of the emerged residuals (rift-shoulder stumps…) and the uplifting regions (Borborema) could provide. We also established the compatibility between the partial erosion of this post-rift cover initiated in post-Cenomanian times and the volumes of sediments accumulated in the offshore basins. No morphological or sedimentological evidence of deep post-Cenomanian burial was detected, except for the Turonian limestones of the Potiguar basin, which were deposited in shallow marine conditions, and for accumulations of Cenozoic clastic sediments (Barreiras Group) along the coast. Even the fluvial sandstones of the Serra do Martins Formation (Paleocene or older?), preserved on isolated mesas at the northern and eastern periphery of the Borborema plateau (Morais Neto et al. 2009), only reach a few tens of metres in thickness and are not convincingly shown to have ever been thicker. Moreover, there is no clear evidence of accelerated offshore sedimentation that might reflect the later erosion of thick post-rift sediments (Peulvast et al. 2008).

As shown higher, the exhumation and partial destruction of paleosurfaces probably began early around the Chapada do Araripe, during the Paleogene. These early stages of topographic inversion might correspond to the first post-rift denudation episode reported by AFT. The second stage, in the Oligocene and later, would correspond to the major phase of river incision and basin inversion. Responding to regional uplift, the differential erosion is also controlled by retreat of the river knickpoints proceeding from the developing low-lying Sertaneja Surface. It reached the lowermost post-rift layers, cutting into the syn-rift formations and even into the underlying basement outcrops of the former rift shoulders (Horto, Serra de Juá, Dom Leme). The corresponding mean denudation rate (7–10 m $Myear^{-1}$) is similar to that of inferred vertical movements.

4.3.4 Eustatic Controls on Erosion and Sedimentation

In the study area, the opening of the Atlantic Ocean coincided with a period of high sea levels (Albian: +160 to +230 m according to Haq et al. 1987; or +50 to +60 m according to Miller et al. 2005), followed by higher mean levels from the Cenomanian to the end of the Maastrichtian. In the onshore Potiguar basin, the continuity of depositional processes during the period of highest sea levels (Albian–Cenomanian to early Campanian) indicates active subsidence until the transient pre-Ubarana uplift event. The subsequent lack of marine deposits suggests that this area remained above sea level as subsidence rates lagged behind rates of sea-level fall during the Cenozoic, exposing the Late Cretaceous units near the coast and allowing some erosional bevelling of stratigraphic packages (Mello 1989). Around the Chapada do Apodi, the absence of Cretaceous deposits on the low erosional plain is the result of stripping of part of the Jandaira limestone and the underlying post-rift layers.

According to Rossetti et al. (2013), the Oligo-Miocene and Miocene corresponded to a time interval of pronounced transgressive events along the equatorial margin of Brazil, and there is a high potential for these episodes to be related to eustatic sea-level rises rather than tectonic subsidence. The same authors show that the Barreiras Formation formed during the early/middle Miocene within transitional coastal settings, during a rising sea level, but before reaching the highstand. The internal stratigraphic units of this formation are related to transgressive-regressive episodes punctuated upon the early/meso Miocene overall rise. Afterwards, a drop in sea level ended the Miocene sedimentation in the Brazilian seaboard, leading to the development of an unconformity and formation of a lateritic soil horizon at the top of the Barreiras Formation. The prevalence of subaerial conditions and lateritic soil formation from mid-Miocene up to late Quaternary is consistent with a prolonged sea-level lowstand, although this stage of soil formation began a few million years before the middle Miocene highstand was reached. This means that even during this supposedly eustatic highstand the Barreiras Formation was already subjected to a prolonged sub-aerial exposure that favored the development of the lateritic profile. This phase would have extended through the late Miocene and even well beyond into the Plio-Pleistocene and Early/Middle Pleistocene (Rossetti et al. 2013). This stage was also that of incipient erosion of the broad apron of Barreiras alluvial fans.

By temporarily putting the sedimentary system out of equilibrium, short-term lowstands (<100 ka) such as those which occurred in the Quaternary, are expected to have increased local erosion more than long-term oscillations (Harris and Mix 2002; Molnar 2004). The dissection of the Barreiras deposits by the main rivers into wide chevrons (Fig. 4.7f), which form the coastal "tabuleiros" or tablelands, also reflects the continuation of the very gentle deformation (tilting and subsidence) which had increased the accommodation space during their deposition (Rossetti et al. 2013; see also Sect. 5.1.1).

4.3.5 Basin Inversion and Possible Mechanisms of Regional Uplift

4.3.5.1 Local Evidence of Basin Inversion? The Araripe Basin

Whereas morphostructural evidence, the nature of stepped surfaces and the scarp sinuosity along the major part of the rims of the Chapada do Araripe plead against the occurrence of local tectonic inversion, for example along branches of the Patos and Pernambuco EW shear zones, the situation is less clear to the NE, above the Cariri depression. Although differential erosion was decisive in the shaping of this depression, several faults have been identified here and in the nearby basement. Some of them may have been reactivated in recent times, as suggested in nearby regions such as the Pereiro massif (Gurgel et al. 2013). Generally, they do not directly deform the post-rift series and the sandstone cap, but one of them, marked by a narrow vertical N160° shear zone, controls a short corridor that separates the Chapada from the thin sandstone cap of the granitic Serra da Boa Vista, NW of Crato. In this case, a post-Cenomanian reactivation is probable, since the base of the sandstone cap is clearly higher on the NE wall than on the opposite side. Both compartments were later bevelled to similar altitudes on the edge of the Chapada.

Other clues for recent tectonic activity were found along a normal fault that cuts the Muriti waste deposit between Crato and Juazeiro do Norte (Peulvast et al. 2011). The vertical throw to the south observed along this WNW-trending fault is 0.5 m and may represent a reactivation of one of the northern boundary faults of the Araripe basin. Fault movement has affected an undated debris apron and records a seismo-tectonic event which may indicate local neotectonic activity. Both structures seem to belong to the same S-shaped fault line, and indicate slight subsidence of the Chapada compartment relative to the slightly uplifted NE border of the basin, which was exhumed from beneath a thinner cover. This fact may have contributed to better preservation of the sandstone cap in the axis of the basin than on its NE border, and therefore to topographic inversion, but it rules out any likelihood of fault-controlled local uplift of the plateau relatively to its surroundings.

In spite of geomorphic evidence such as the preservation of Exu sandstone outliers at similar altitudes outside the Araripe basin, to the southwest (Serra Vermelha) and far to the south, on the north rim of the Jatoba basin, the Chapada do Araripe was recently interpreted as the result of inversion of an intracontinental graben to a high-standing horst (Marques et al. 2014). This extreme proposition relies on the wrong assumption that the top sandstones "are not well consolidated and, thus, are friable, which makes them little resistant…", and would lead to "expect to see them in depression relative to resistant host basement" (ibid, p. 257). This interpretation is contrary to multiple evidence of high resistance of the sandstone cover relatively to the basement and to the rest of its sedimentary cover, as demonstrated higher.

Tectonic arguments are mainly based upon poor field evidence of thrust movements along faults that are inferred from the DEM but "not clearly recognized in the field" (ibid, p. 251). These faults, mainly the NW–SE and ENE–WSW ones, would have been reactivated by compression of the basin between the EW Patos and Pernambuco lineaments, in a restraining bend of the left-lateral shear zone system, in the context of horizontal compressive stress at continental scale, the Mid-Atlantic Ridge-push from the east and the Andan-push from the west. The total inversion would be 2500 m or more. Beside this highly improbable estimation based on thermochonological data, (Sect. 4.3.3), the most critical point is the identification of reverse faults, on hypothetic profiles that do not match with field reality (the so-called stair-stepping topography of the chapada), or on small outcrops in the Nova Olinda region, located in landslide deposits or in the basal units of the post-rift cover where the "overthrust" sandstones are not those of the top cover, but those of lower layers (probably Batateira sandstones). According to the cartoon published by these authors (Marques et al. 2014, their Fig. 18), slices of Exu sandstone would be overthrust on the borders of the inferred horst and preserved from erosion in the north and south piedmonts. This is contrary to field evidence and even to the geological map reproduced by the authors in the same article.

Better evidence of basin inversion has been presented by the same authors for the Rio do Peixe basins (Nogueira et al. 2015), corresponding to the same shift in regional stress field, in Early Miocene. However, much weaker effects are recognized in these basins where no remnant of post-rift cover is identified, and no significant geomorphic consequences (fault scarps, etc.) are reported, as shown in our study of differential erosion landforms (Sect. 3.1.1).

In our opinion, the elevated position of the Chapada do Araripe corresponds to a tectonic inversion of regional proportions, better termed topographic inversion. In the whole region, post-Cenomanian uplift caused an inversion of the Cretaceous basins and generated a landscape in which the most elevated landforms correspond either to resistant Mesozoic sedimentary caprock or to eroded stumps of syn-rift Cretaceous footwall uplands (Peulvast et al. 2008). This post-Cenomanian crustal deformation mimics the geometry of a broad monocline between the Ceará coast, to the north, and the Chapada do Araripe, with a half wavelength of ~300 km (Fig. 4.6). The Araripe basin was the most vigorously uplifted portion of the landscape. The ~650 m of post-Albian crustal uplift in the area now forming a major drainage divide ca 300 km from the coast also estimate minimum long-term surface uplift of this part of the Brazilian shield. The flexural style of this uplift explains the elevated position of the former Cenomanian fluvial plain relatively to the northern plateaus, where the former basin edge was slightly tilted northward and then destroyed by subsequent stages of denudation and differential erosion. This inferred epeirogeny is partly related, at least in its latest stages, to Cenozoic uplift of the Borborema plateau.

4.3.5.2 Regional Basin Inversion

Passive margin escarpment models have tended to dwell on the hypothesis that present day escarpments at continental margins are direct descendants of a single tectonic event in the geologic past—namely continental breakup. For that reason, models have explored a range of theoretical mechanisms that were capable of either generating escarpments by headward erosion into the flank of a syn-rift seaward flexure (Fig. 4.9a), or making eroding fault scarps persist indefinitely through time by upward flexural rebound (e.g., King 1955; Gilchrist and Summerfield 1990 and Fig. 4.9b). This study supports instead a growing consensus based on investigations in southeastern Australia (Cockburn et al. 2000; Persano et al. 2002), Africa generally (Burke and Gunnell 2008), and India (Gunnell et al. 2003) that rift shoulders erode relatively rapidly (i.e. within <40 Myear) after their formation, even though locally resistant rock outcrops that were once part of the rift shoulder may survive much longer as residual landforms (cf. the Ibiapaba–Baturité and Aracati–Macau alignments). In NE Brazil, where Cretaceous basins such as the Araripe currently form elevated plateaus, uplift has been caused by some much later tectonic overprint, unrelated to breakup dynamics. Resulting seaward-facing escarpments mimic rifted margin escarpments due to their location close to oceanic basins, but their genetic link with continental breakup is deceptive. This conceptual model (Fig. 4.9c) involving two separate episodes of crustal deformation, i.e. rift-flank uplift and erosion followed by swell uplift and swell-flank erosion ~60 Myear

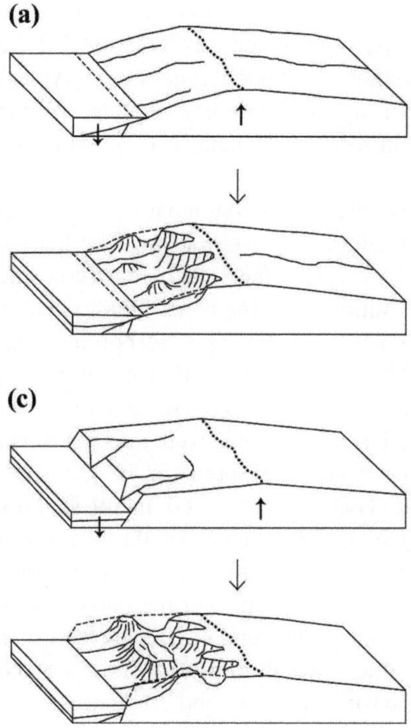

Fig. 4.9 Conceptual models of landscape development at passive margins. **a** Downwarping model. Dissection of a monocline formed during breakup, where scarp retreat leaves trailing remnants of the downwarped plateau. Stability of the drainage divide pinned at the inland edge of the downwarped region. **b** Flank uplift model, involving faulting, erosional retreat of the scarp and of the drainage divide. Initial topography dips away from margin. **c** Inclined plateau model. Initial topography consists of a gently seaward-dipping plateau, seaward of a pre-breakup drainage divide situated in the hinterland. No further uplift occurs during breakup, but base level drops to sea level. Establishment of a secondary drainage divide developing into an escarpment as a result of isostatic rebound seaward of the drainage divide. The escarpment is established and maintained at the locus of maximum isostatic rebound and does not retreat. Compiled and adapted after Van der Beek and Braun (1999), Ollier and Pain (2000), and authors' observations in Ceará

later, is where the interpretation of the regional topographic staircase in this book differs most fundamentally from previous views in the litterature. The main question arising from this reinterpretation lies with finding the causes of Cenozoic uplift and basin inversion.

The causes of basin inversion and the exact timing of uplift are currently speculative. Harman et al. (1998) suggested that the increase in average rate of denudation recorded a late Cretaceous event probably related to the uplift of the eastern margin of Brazil at 60–80 Myear as a response to major changes in the relative plate motions between Africa, Antarctica, and South America. A similar explanation of regional uplift and basin inversion on the Brazilian shield (Lima et al. 2002) has called upon tectonic buckling caused by E–W compression throughout South America in relation to the Andean orogeny since the Oligocene (Gregory-Wodzicki 2000). This could provide a more general explanation for the existence of the 'chapadas', or elevated sedimentary tablelands, that are so typical of Brazilian cratonic scenery (Arai 2000). Such an explanation, completed by the role or Mid-Atlantic Ridge-push, is favoured in recent publications (e.g., Marques et al. 2014).

Also recognized in the Recôncavo–Tucano–Jatobá area (Magnavita et al. 1994), the main regional uplift stage coincided with Oligocene magmatism. The known surface manifestations of this magmatism are limited to a few small areas (volcanic plugs and lava flows: Morais Neto et al. 2002), remote from the zones identified here of both minimal erosion and maximum uplift (Chapada do Araripe, Borborema, etc.: Fig. 4.7e). In Africa, Burke and Gunnell (2008) have argued that a number of seaward-facing escarpments around the continent were geologically younger than 30 Myear and often associated with volcanism. The escarpments were thus not direct descendants of rift shoulders formed during continental breakup in the early Mesozoic, but corresponded instead to the eroded oceanic limbs of dynamic shallow mantle swells that had formed more recently as a consequence of the prolonged stationarity of the continent relative to the Earth's mantle. Although South America is not known to share the unique tectonic history and current thermal buoyancy of the African plate, the influence of a Cenozoic magmatic underplate (instead of a thermal swell) is possible. This would have thickened the base of the crust and thereby generated regional crustal updoming, basin inversion, and erosion.

This is the hypothesis proposed by Oliveira and Medeiros (2012), according to which the largest portion of the Cenozoic magmatism would have been retained at the base of the crust, forming magmatic underplating, and/or intruded in the original lower crust, while a minor portion would have reached the Earth surface. In this context, the crustal thickening under the Borborema Plateau (Fig. 4.10) could be attributed to the added material at the base of the crust. The negative density contrast between the underplated magmatic rocks and the surrounding mantle would have produced a buoyant force that would have deformed the crust and uplifted the region, thus forming the Borborema Plateau Facts that are consistent with the hypothesis of magmatic underplating are the crustal thickening of about 4 km below the Borborema Plateau (De Castro et al. 1997, 1998) and the existence of an important discontinuity in the intracrustal seismic velocity associated with high Vp/Vs ratio (Barbosa 2008). According to Musacchio et al. (1997), high Vp/Vs ratios are clues of basic composition for the crust.

However, some facts are not consistent with the hypothesis of magmatic underplating, such as the low rock volume of the Cenozoic magmatism, and the nature of this magmatism that is constituted mainly of alkaline rocks. According to Lima et al. (2015) and their seismic profile, the actual thickness and internal structure of the crust of the Borborema province show that it was affected by traction processes that strongly reworked and thinned the crust, a process possibly related to the breakup of Pangaea in the Cretaceous, leading to the South Atlantic Ocean opening. There would be no seismic evidence of crustal underplating under the seismic profile. Seismic data obtained in this study, mainly low crustal Vp, upward bending

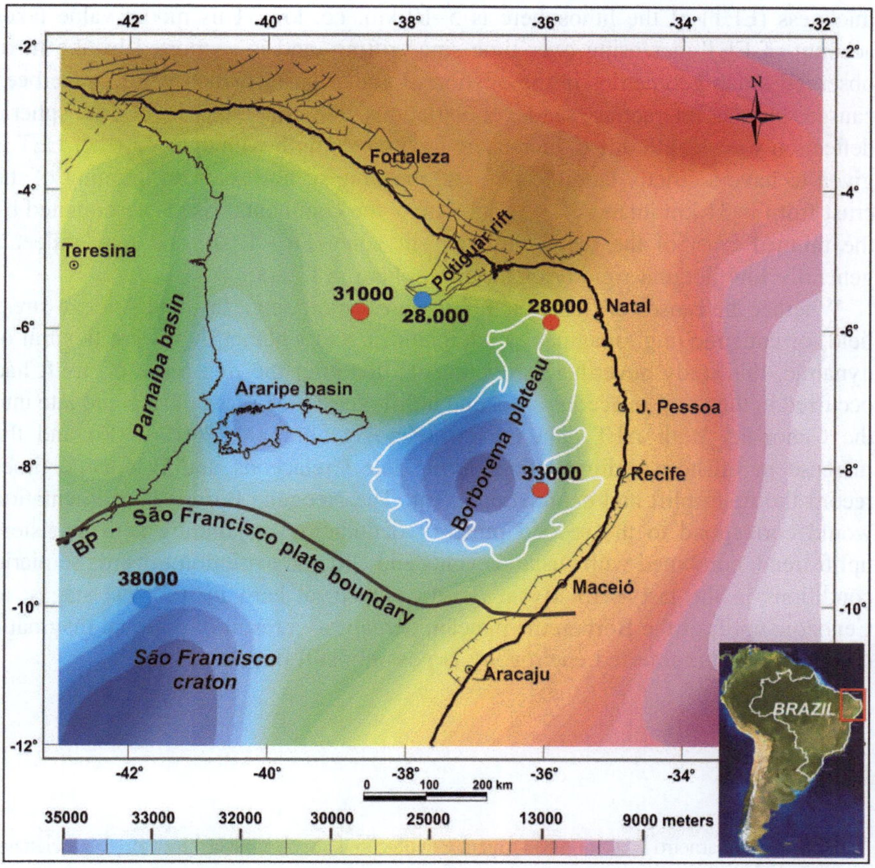

Fig. 4.10 Crust thickness estimates for the Borborema Province and adjacent areas. Shaded-relief map (Inc = 35° and Dec = 315° Az). The boundary of the Borborema Province is indicated in *solid black line*. *Blue* and *red dots* show points where estimates of the crust thickness are known. From Oliveira and Meideros (2012)

of upperlower crust limit, smooth Moho topography, and low lithospheric mantle velocity, strongly suggest hot mantle below the Borborema Plateau, and that uplift of the plateau was caused by delamination and removal of mafic material from the base of the crust.

Finally, because the lithosphere deflects if a load is applied to the Earth's surface, we also expect a contribution to flexural deformation from denudational offloading on shore associated with sediment loading of the continental margin (Bittencourt et al. 1999; Petit et al. 2007). Here the contribution from this process is potentially limited because of the generally low rates of Cenozoic sediment loading, excepted in the offshore Potiguar depocentre. However, isostatic response functions obtained for the offshore margin based on bathymetry and free-air gravity anomalies (Mello and Bender 1988) have shown that the time-averaged effective elastic thickness (EET) of the lithosphere is 5–10 km, i.e. low. This model value takes account of EET increasing over time since rifting, and is supported by the fit of observed onlap geometries in the offshore Potiguar basin, which would have been caused by the interacting sea-level variations and the increase of lithospheric deflection wavelength due to increasing flexural strength with age. Whether EET is given to have a strictly tangible physical meaning or not, seaward thinning of the crust from ~31 km inland to ~10 km below the continental rise is accentuated by the thinned crust of the failed Potiguar rift zone (Fig. 4.10), and may reflect a generally low flexural rigidity of the lithosphere in Ceará.

Whether the causes of Cenozoic uplift were remote (e.g., far-field Andean stress field) or endemic (e.g., magmatic underplating), and whether they were flexural or dynamic, this study nevertheless documents that after the first pulse of uplift had occurred in the late Cretaceous, a second uplift phase took place during and late into the Cenozoic. Both AFT data (Morais Neto et al. 2000, 2005–2006) and the increase in offshore sedimentation during late Cretaceous and Paleogene times record the first uplift and denudation event. The Neogene Barreiras sedimentation would correspond to the second pulse of denudation responding to a persistent uplift trend, combined with eustatic events and climatic evolution towards semiarid conditions in the last ~15 Myear. Partly related, at least in its latest stages, to Cenozoic uplift of the Borborema plateau, perhaps as a result of regional magmatic underplating, this inferred epeirogeny is possibly still ongoing.

References

Almeida FFM, Carneiro CDR (1998) Origem e evolução da Serra do Mar. Rev Bras Geociências 28(2):135–150

Andrade JAFG, Brito PMB, Amaral CR, Machado LP, Mayrinck D, Machado MS (2002) Paleobiota do Membro Crato da Formação Santana. XLI Congr Bras Geol, João Pessoa, Anais, Cretáceo inferior (Aptiano) da Bacia do Araripe, p 653

Arai M (1999) A transgressão marinha mesocretácea: sua implicação no paradigma da reconstitução paleogeográfica do Cretáceo no Brasil. Bol 5° Simp. sobre o Cretáceo do Brasil, UNESP, Rio Claro, 577–582

Arai M (2000) Chapadas: relict of mid-Cretaceous interior seas in Brazil. Rev Bras Geociências 30 (3):436–438

Araripe PT, Feijo FJ (1994) Bacia Potiguar. Bol Geoci Petrobras 8(1):127–141

Archanjo CJ, Trindade RI, Macedo JWP, Araújo G (2000) Magnetic fabric of a basaltic dyke swarm associated with Mesozoic rifting in northeast Brazil. J South Am Earth Sci 13:179–189

Assine ML (1994) Paleocorrentes e paleogeografia na Bacia do Araripe. Nordeste do Brasil. Rev Bras Geociências 24(4):223–232

Barbosa MFN (2008) Estimativa deEspessura Crustal na ProvínciaBorborema (NE/Brasil) através defunção Receptor. Msc Thesis, Universidade Federal do RioGrande do Norte, Natal

Baudin F, Berthou PY (1996) Environnement de dépôt de la matière organique des sédiments aptiens-albiens du basin d'Araripe (NE du Brésil). Bull Centres Rech Explor Prod Elf-Aquitaine 20(1):213–227

Beltrami CV, Alves LEM, Feijo F (1994) Bacia do Ceará. Bol Geoci Petrobras 8(1):117–125

Bétard F (2007) Montagnes humides au cœur du Nordeste brésilien semi-aride: le cas du massif de Baturité (Ceará). Apports d'une approche morphopédologique pour la connaissance et la gestion des milieux. Ph.D thesis, Univ Paris-Sorbonne

Bétard F (2012) Spatial variations of soil weathering processes in a tropical mountain environment: the Baturité massif and its piedmont (Ceará, NE Brazil). Catena 93 (18–28):2012. doi:10.1016/j.catena.01.013

Bétard F, Caner L, Gunnell Y, Bourgeon G (2009) Illite neoformation in plagioclase during weathering: evidence from semi-arid Northeast Brazil. Geoderma 152:53–62. doi:10.1016/j.geoderma.05.016

Bétard F, Peulvast JP (2013) Non-lateritic red soils of the Brazilian drylands: pedogenic processes and landscape evolution. In: 8th international conference on geomorphology, Paris, 27–31 Aug 2013, Abstracts volume, p 594

Bétard F, Peulvast JP, Claudino Sales V (2005) Laterite preservation and soil distribution in the Araripe-Campos Sales area, Northeastern Brazil: consequences of uplift, erosion and climatic change. In: 6th international conference on geomorphology, Zaragoza, 7–11 Sept 2005, Abstracts Volume, p 69

Bezerra FHR, Amaro VE, Vita-Finzi C, Saadi A (2001) Pliocene-Quaternary fault control of sedimentation and coastal plain morphology in NE Brazil. J South Am Earth Sci 14:61–75

Bierman PR, Caffee M (2001) Slow rates of rock surface erosion and sediment production across the Namib desert and escarpment, southern Africa. Am J Sci 301:326–358

Bittencourt ACSP, Dominguez JML, Ussami N (1999) Flexure as a tectonic control on the large scale geomorphic characteristics of the eastern Brazil coastal zone. J Coastal Res 15 (2):505–519

Boillot G, Coulon C (1998) La déchirure continentale et l'ouverture océanique. Géologie des marges passives. Gordon and Breach Sci Publ, Amsterdam

Brunsden D (1993a) The persistance of landforms. Z Geomorph, Suppl Bd 93:13–28

Brunsden D (1993b) Barriers to geomorphological change. In: Thomas DSG, Allison RJ (eds) Landscape sensitivity. Wiley, Chichester, pp 7–12

Burke K, Gunnell Y (2008) The African erosion surface: a continental-scale synthesis of geomorphology, tectonics and environmental change over the past 180 million years. Geol Soc Am Memoir 201:1–66

Carneiro CDR, Hamza VM, Almeida FFM (1989) Ativação tectônica, fluxo geotérmico e sismicidade no Nordeste oriental brasileiro. Rev Bras Geociências 19(3):310–322

Castro JC, Barrocas SLS, Lima HP (1981) Fácies, diagênese e modelos de acumulação da Formação Açu, parte emersa da Bacia Potiguar. Petrobras, Cenpes

Cockburn HAP, Brown RW, Summerfield MA, Seidl MA (2000) Quantifying passive margin denudation and landscape development using a combined fission track thermochronology and cosmogenic isotope analysis approach. Earth Planet Sci Lett 179:429–435

Condé VC, Lana CC, Pessoa Neto OC, Roesner EH, Morais Neto JM, Dutra DC (2007) Bacia do Ceará. Bol Geoci Petrobras 15(2):347–355

Costa IG, Beltrami CV, Alves LEM (1990) A evolução tectono-sedimentar e o habitat do óleo da Bacia do Ceará. Bol Geoci Petrobras 4(1):65–74

CPRM-Serviço Geológico do Brasil (2001) Geologia, tectônica e recursos minerais do Brasil. Sistema de Informações Geográficas – SIG e mapas na escala 1:2,500,000. Serviço Geológico do Brasil, 4 CD Rom.

CPRM-Serviço Geológico do Brasil (2003) Atlas digital de geologia e recursos minerais do Ceará. Mapas na escala 1:500,000. Serviço Geológico do Brasil, CD Rom

Cunha AAS (1991) Reconhecimento dos hiatos da seção terciária da sub-bacia de Mundaú (Bacia do Ceará): um estudo baseado em nanofosseís calcários. Bol Geoci Petrobras 8(1):69–82

Cunha LS, Castelo Branco RMG (2002) Configuração 3-D dos sedimentos quaternaries baseada em inverses de dados geofísicos elétricos e eletromagnéticos, região costeira de Cascavel – Ceará. XLI Congr Bras Geol, João Pessoa, Anais, p 85

De Castro DL, Barbosa VCF, Silva JBC, Medeiros WE (1997) Relevo da interface crosta-manto entre os vínculos de isostasia e suavidade. Congr Int SBGf 5, São Paulo, Anais, pp 682–685

De Castro DL, Medeiros WE, Jardim de Sá EF, Moreira JAM (1998) Gravimetric map of the northeastern Brazil and adjacent continental margin: interpretation based on the isostatic hypothesis. Revista Brasileira de Geophysica 16:115–131

Gardner TW, Braun DD, Pazzaglia FJ, Sevon WD (1993) Late Cainozoic landscape evolution of the Susquehanna River basin. In: 3rd international geomorphology conference Hamilton, post-conference field trip guidebook, Aug 1993

Gilchrist AR, Summerfield MA (1990) Differential denudation and flexural isostasy in formation of rifted margin upwarps. Nature 346:739–742

Gregory-Wodzicki KM (2000) Uplift history of the Central and Northern Andes: a review. Geol Soc Am Bull 112:1091–1105

Gunnell Y (2003) Radiometric ages of laterites and constraints on long-term denudation rates in West Africa. Geology 31:131–134

Gunnell Y, Gallagher K, Carter A, Widdowson M, Hurford AJ (2003) Denudation history of the continental margin of western peninsular India during the Mesozoic and Cenozoic. Earth Planet Sci Lett 215:187–201

Gurgel SPP, Bezerra FHR, Corréa ACB, Marques FO, Maia RP (2013) Cenozoic uplift and erosion of structural landforms in NE Brazil. Geomorphology 186:68–84

Haq BU, Hardenbol J, Vail PR (1987) The chronology of fluctuating sea level since the Triassic. Science 235:1156–1167

Harman R, Gallagher K, Brown R, Raza A (1998) Accelerated denudation and tectonic/geomorphic reactivation of the cratons of northeastern Brazil during the Late Cretaceous. J Geophys Res 103:27091–27105

Harris SE, Mix AC (2002) Climate and tectonic influences on continental erosion of tropical South America, 0–13 Ma. Geology 30(5):447–450

Hegarty KA, Morais Neto JM, Karner GD (2002) The enigma of the Araripe plateau: new constraints on its uplift and tectonic history using AFTA. XLI Congr Bras Geol, Joã Pessoa, Anais, p 495

Huber BT, Norris RD, MacLeod KG (2002) Deep-sea paleotemperature record of extreme warmth during the Cretaceous. Geology 30(2):123–126

Japsen P, Bonow JM, Green PF, Cobbold PR, Chiossi D, Lilletveit R, Magnavita LP, Pedreira A (2012) Episodic burial and exhumation in NE Brazil after opening of the South Atlantic. Geol Soc Am Bull 124:800–816

Jardim de Sá EF, Matos RMD, Morais Neto JM, Saadi A, Pessoa Neto OC (1999) Epirogenia Cenozózoica na Província Borborema: síntese e discussão sobre os modelos de deformação associados. VII Simp. Nac. Estudos Tect, Lençois – Bahia, Anais, pp 58–61

Jelinek AR, Chemale F Jr, van der Beek PA, Guadagnin F, Cupertino JA, Viana A (2014) Denudation history and landscape evolution of the northern East-Brazilian continental margin from apatite fission-track thermochronology. J S Am Earth Sci 54:158–181

King LC (1955) Pediplanation and isostasy: an example from South Africa. Quart J Geol Soc London 111:353–359

Krook L (1975) Surinam. In: Fairbridge RW (ed) The Enclopaedia of world regional geology, Part 1. Western Hemisphere (Including Antarctica and Australia), Dowden, Hutchinson & Ross, Stroudsburg, Pa, pp 480–492

Kroonenberg SB, Melitz PJ (1983) Summit levels, bedrock control and the etchplain concept in the basement of Suriname. Geol en Mijnbouw 62:389–399

Lageat Y, Gunnell Y (2001) Landscape development in tropical shield environment. In: Godard A, Lagasquie JJ, Lageat Y (eds) Basement regions. Springer, Berlin, pp 173–197

Lanson B, Sakharov BA, Claret F, Drits VA (2009) Diagenetic smectite-to-illite transition in clay-rich sediments: A reappraisal of X-ray diffraction results using the multi-specimen method. Am J Sci 309(6):476–516

Lappartient JR (1985) Le Continental Terminal et le Pléistocène ancien du bassin sénégalo-mauritanien : stratigraphie, sédimentation, diagenèse, altération; reconstitution des paléorivages au travers des cuirasses. Ph.D thesis, Univ Aix-Marseille 3

Lima CC, Silva AT, Guerra MCM, Pequeno MA, Lopes MRF, Roberto SAR, Quental SHAJ, Machado Jr DL (2002) Modelos físicos da inversão incipiente de bacias da margem sudeste brasileira. XLI Congr Bras Geol, João Pessoa, Anais, p 270

Lima MVAG, Berrocal J, Soares JEP, Fuck RA (2015) Deep seismic refraction experiment in northeast Brazil: New constraints for Borborema province evolution. J S Am Earth Sci 58: 335–349. doi:10.1016/j.jsames.2014.10.007

Łuszczak K, Persano C, Stuart F (2014) Low temperature thermochronological constraints on the Cenozoic evolution of the Lake District and Southern Uplands Massifs (NW England, SW Scotland). 14th International conference on thermochronology, Chamonix, France

Magnavita LP, Davison I, Kuznir NJ (1994) Rifting, erosion and uplift history of the Reconcavo-Tucano-Jatoba Rift, northeast Brazil. Tectonics 13:367–388

Marques FO, Nogueira FCC, Bezerra FHR, de Castro DL (2014) The Araripe Basin in NE Brazil: an intracontinental graben inverted to a high-standing horst. Tectonophysics 630:251–264

Martill DM (1993) Fossils of the Santana and Crato Formations, Brazil (Field Guide to Fossils no. 5). The Palaeontological Association, London

Matos RMD (1992) The Northeast Brazilian rift system. Tectonics 11(4):766–791

Matos RMD (2000) Tectonic evolution of the Equatorial South Atlantic. In: Mohriak W, Taiwani M (eds) Atlantic Rifts and continental margins. Geophys Monograph 115, Am Geophys Union, pp 331–354

Mello UT (1989) Controles tectônicos na estratigrafia da Bacia Potiguar: uma integração de modelos geodinâmicos. Bol Geoci Petrobras 3(4):347–364

Mello UT, Bender AA (1988) On isostasy at the equatorial margin of Brazil. Revista Brasileira de Geociências 18:237–246

Michel P (1973) Bassins des fleuves Sénégal et Gambie. Etude géomorphologique. Mém ORSTOM 63, Paris

Miller KG, Kominz MA, Browning JV, Wright JD, Mountain GS, Katz ME, Sugarman PJ, Cramer BS, Christie-Blick N, Pekar SF (2005) The Phanerozoic record of global sea-level change. Science 310:1293–1298

Molnar P (2004) Late Cenozoic increase in accumulation rates of terrestrial sediment: how might climate change have affected erosion rates? Annu Rev Earth Planet Sci 32:67–89

Morais Neto JM (1999) As coberturas sedimentares tertiarias do interior da Paraiba e do Rio Grande do Norte e gênese da anteclise da Borborema. MSc thesis, Univ Federal de Ouro Preto, Ouro Preto, Brazil

Morais Neto JM, Hegarty K, Karner GD (2005–2006) Abordagem preliminar sobre paleotemperatura e evolução do relevo da bacia do Araripe, Nordeste do Brasil, a partir da análise de traços de fissão em apatita. Bol Geoci Petrobras 14:113–119

Morais Neto JM, Hegarty KA, Karner GD, Alkmim FF (2009) Timing and mechanisms for the generation and modification of the anomalous topography of the Borborema Province, northeastern Brazil. Mar Pet Geol 26:1070–1086

Morais Neto JM, Hegarty KA, Karner GD, Alkmim FF, Matos RMD (2000) Uplift and erosion in the Borborema Province, northeast Brazil: insights from apatite fission track analysis. In: XXXI International Geology Congress, IUGS/SBG, Abstract volume, Rio de Janeiro

Morais Neto JM, Menezes MRF, Szatmari P, York D (2002) Datação Ar/Ar do plug basáltico «Serrote Preto» e seu significado para a cronologia de Formação Serra do Martins. In: XLI Congress Brasil Geology, João Pessoa, Anais, p 499

Morais Neto JM, Vasconcelos PM, Stone J (2010a) Tuning the landscape evolution in the Borborema Province, northeastern Brazil, by reconciling low-temperature thermochronology and ^{10}Be isotope analysis. In: VII SSAGI-South American symposium on isotope geology, Brasilia

Morais Neto JM, Vasconcelos PM, Stone J, Lima MG (2010b) Variability of erosion rates captured by ^{10}Be cosmogenic isotope: a case study in the Seridó valley, northeastern Brazil. In: VII SSAGI-South American symposium on isotope geology, Brasilia

Musacchio G, Mooney WD, Luetgert JH, Christensen NI (1997) Composition of the crust in the Grenville and Appalachian Provinces of North America inferred from Vp/Vs ratios. J Geophys Res Solid Earth 102(B7):15225–15241

Neto JA, Mort H, Pereira R, Barbosa J, Neumann V, Vortisch W, Filho OJC, Brandão PALS, Pacheco JG (2013) Carbonaceous shales in the Araripe Basin, NE Brazil: a potential shale gas reservoir. Search and Discovery Article #80309, http://www.searchanddiscovery.com/pdfz/documents/2013/80309neto/ndx_neto.pdf.html

Neumann VH (1999) Estratigrafía, sedimentología, geoquímica y diagénesis de los sistemas lacustres aptiense-albienses de la Cuenca de Araripe (Noreste de Brasil). Ph.D thesis, Univ Barcelona

Neves CAO (1989) Analise regional do trinomio geração-migração-accumulação de hidrocarbonetos na seqüência continental eocretácica da Bacia Potiguar. Bol Geoci Petrobras 3(3):131–145

Nogueira FCC, Marques FO, Bezerra FHR, De Castro DL, Fuck RA (2015) Cretaceous intracontinental rifting and post-rift inversion in NE Brazil: insights from the Rio do Peixe basin. Tectonophysics 644–645:92–107. doi:10.1016/j.tecto.2014.12.016

Oliveira DC, Gomes CJS (1996) A expressão de reativações pós-intrusão do enxame de diques Rio Ceará Mirim (Mesozóico): implicações na evolução tectônica da Bacia Potiguar. Bol Geoci Petrobras 10(1–4):121–145

Oliveira RG, Medeiros WE (2012) Evidences of buried loads in the base of the crust of Borborema Plateau (NE Brazil) from Bouguer admittance estimates. J S Am Earth Sci 37:60–76

Ollier CD, Pain CF (1997) Equating the basal unconformity with the palaeoplain: a model for passive margins. Geomorphology 19:1–15

Ollier CD, Pain CF (2000) The Origin of Mountains. Routledge, London

Persano C, Stuart FM, Bishop P, Barford DN (2002) Apatite (U-Th)/He age constraints on the development of the great escarpment on the southeastern Australian passive margin. Earth Planet Sci Lett 200:79–90

Pessoa Neto OC (1999) Análise estratigráfica integrada da plataforma mista (siliciclástica-carbonática) do Neogeno da Bacia Potiguar, nordeste do Brasil. MSc thesis, Univ Fed Rio Grande do Sul, Porto Alegre, Brazil

Pessoa Neto OC (2003) Estratigrafia de seqüências da plataforma mista neogênica na Bacia Potiguar, margem equatorial brasileira. Rev Bras Geociências 33:263–278

Petit C, Fournier M, Gunnell Y (2007) Tectonic and climatic controls on rift escarpments: erosion and flexural rebound of the Dhofar passive margin (Gulf of Aden, Oman). J Geophys Res 112 (B03406). doi:10.1029/2006JB004554

Petri S (1987) Cretaceous paleogeographic maps of Brazil. Palaeogeogr Palaeoclim Palaeoecol 59:117–168

Peulvast JP, Bétard F (2013) Late Cenozoic and present-day hillslope erosion dynamics in a passive margin context: stability or instability? Case studies in Northeast Brazil. Geografia Fisica e Dinamica Quaternaria 36(1):139–149. doi:10.4461/GFDQ.2013.36.0

Peulvast JP, Bétard F (2015) A history of basin inversion, scarp retreat and shallow denudation: the Araripe Basin as a keystone for understanding long-term landscape evolution in NE Brazil. Geomorphology 233:20–40. doi:10.1016/j.geomorph.2014.10.009

Peulvast JP, Bétard F, Lageat Y (2009) Long-term landscape evolution and denudation rates in shield and platform areas: a morphostratigraphic approach. Géomorphologie: relief, processus, environnement 2:95–108. doi:10.4000/geomorphologie.7540

Peulvast JP, Bétard F, Magalhães AO (2011) Scarp morphology and identification of large-scale mass movements in tropical tablelands: the eastern Araripe basin (Ceará, Brazil). Géomorphologie: relief, processus, environnement 1:33–52. doi:10.4000/geomorphologie.8800

Peulvast JP, Claudino Sales V (2003) Carta morfoestrutural do Ceará e áreas adjacentes do rio Grande do Norte e da Paraíba, escala 1 :1 000.000. In: Atlas Digital de Geologia e Recursos Minerais do Ceará, CPRM (Serviço Geológico do Brasil)

Peulvast JP, Claudino Sales V (2004) Stepped surfaces and palaeolandforms in the northern Brazilian "Nordeste": constraints on models of morphotectonic evolution. Geomorphology 62:89–122

Peulvast JP, Claudino Sales V, Bétard F, Gunnell Y (2008) Low post-Cenomanian denudation depths across the Brazilian Northeast: implications for long-term landscape evolution at a transform continental margin. Global Planet Change 62:39–60. doi:10.1016/j.gloplacha.2007.11.005

Ponte FC, Ponte-Filho FC (1996) Estrutura geológica e evolução tectônica da Bacia do Araripe. MME/DNPM, Recife

Ricordel-Prognon C, Lagroix F, Moreau MG, Thiry M (2010) Lateritic paleoweathering profiles in French Massif Central: Paleomagnetic datings. J Geophys Res 115(B10104). doi:10.1029/2010JB007419

Rossetti DF, Bezerra FHR, Dominguez JLR (2013) Late Oligocene-Miocene transgressions along the equatorial and eastern margins of Brazil. Earth-Sci Rev 123:87–112

Schmitt JM (1999) Weathering, rainwater and atmosphere chemistry: example and modelling of granite weathering in present conditions and in a CO_2 rich, and in an anoxic atmosphere. Spec Publ Int Ass Sediment 27:21–41

Shimabukuro S, Arai M (2001) A discordância tortoniana e sua relação com o arcabouço estratigráfico do Grupo Barreiras e unidades correlatas (Neogeno do Brasil). In: XVII Congress Brasil Paleontologia, Rio Branco, Acre, Boletim de resumos, p 54

Sousa DC, Jardim De Sá EF, Matos RMD, Oliveira DC (1999) Deformação sin- e pós-Formação Barreiras na região de Ponta Grossa (Icapuí, CE), litoral ocidental da Bacia Potiguar. In: VII Simp Nac Estudos Tect, Lençoís – Bahia, Anais, pp 90–93

Sousa DC, Jardim de Sá EF, Antunes AF (2008a) Deformação neógena e suas implicações na estruturação dos campos de petróleo na região de Icapuí-Ponta Grossa (CE), Bacia Potiguar emersa. Revista Brasileira de Geologia 38(2-suplemento):97–110

Sousa DC, Jardim de Sá EF, Vital H, do Nascimento MAL (2008b) Falésias na praia de Ponta Grossa, Icapuí, CE. Importantes deformações tectônicas cenozóicas em rochas sedimentares da Formação Barreiras. In: Winge M, Schobbenhaus C, Souza CRG, Fernandes ACS, Berbert-Born M, Queiroz ET (eds) Sítios geológicos e paleontológicos do Brasil. Published in Internet 24 Sept 2008. http://www.und.br/ig/sigep/sitio120.pdf

Stone JO, Vasconcelos PM (1999) Quaternary erosion rate estimates from ancient Australian cratonic surfaces. Regional Conference on Geomorphology, Rio de Janeiro, abstract, p 83

Tardy Y, Roquin C (1998) Dérive des continents, paléoclimats et altérations tropicales. Ed BRGM, Orléans

Thiry M, Schmitt JM, Simon-Coinçon R (1999) Problems, progress and future research concerning palaeoweathering and paleosurfaces. In: Thiry M, Simon-Coinçon R (eds) Palaeoweathering, palaeosurfaces and related continental deposits. Spec Publ Int Assoc Sedim 27, pp 3–17

Thomas MF (1980) Timescales of landform development on tropical shields—A study from Sierre Leone. In: Cullingford RA, Davidson DA, Lewin J (eds) Timescales in Geomorphology. Wiley, Chichester, pp 333–354

Van der Beek P, Braun J (1999) Controls on post-mid-Cretaceous landscape evolution in the southeastern highlands of Australia: insights from numerical surface process models. J. Geophys. Res. 104:4945–4966

Van der Wateren FM, Dunai TJ (2001) Late Neogene passive margin denudation history—cosmogenic isotope measurements from the central Namib desert. Global Planet Change 30:271–307

Watchman AL, Twidale CR (2002) Relative and "absolute" dating of land surfaces. Earth-Sci Rev 58:1–49. doi:10.1016/S0012-8252(01)00080-0

Widdowson M, Gunnell Y (1999) Lateritization, geomorphology and geodynamics of a passive continental margin : the Konkan and Kanara coastal lowlands of western peninsular India. In: Thiry M, Simon-Coinçon R (eds) Palaeoweathering, palaeosurfaces and related continental deposits. Spec Publ Int Assoc Sedim 27, pp 245–274

Wildman M, Brown R, Persano C, Stuart F, Beucher R (2014) Intracontinental deformation of the interior plateau and Atlantic continental margin of South Africa: insights from combined AFT and AHe analysis. In: 14th International conference on thermochronology, Chamonix, France

Zonneveld JIS (1985) Geomorphological notes on the continental border in the Guyanas (N. South America). Z Geomorph NF, Suppl Bd 54:71–83

Chapter 5
Late Cenozoic Evolution of Landforms: The Influence of Neotectonics and Climatic Changes

Abstract Forming a continuous clastic apron near the coast, the mainly Neogene Barreiras Formation records the last peak of erosion in the hinterland, which coincided with the onset of more arid climates at ~13 Myear or earlier. Dissected pediments and sparse debris fans visible at the base of some escarpments suggest recent erosive activity and possible slope retreat during the Late Cenozoic. A study of landforms located near zones of seismotectonic activity indicates a possible, probably weak, contribution of neotectonics to the formation of a few high scarps. Many of them are mainly inherited landforms, initiated during the Early Cretaceous rifting or the later margin uplift, a situation which suggests long-term stability. Later on, except in high inner tablelands (Chapada do Araripe), only slight or local backwearing took place, associated with downwearing on low surfaces and pediments, probably in diachronic ways. The Neogene clastic sedimentation on piedmonts and coastal areas mainly reflects the occurrence of dry periods inducing widespread stripping of deep soil horizons and erosion of bare rock slopes and surfaces. Dissection stages occurred in periods of more humid climate and/or low sea level. Marks of strong recent or present activity are mainly registered in the rims of the Chapada do Araripe, owing to favorable structural, hydrogeological and climatic conditions. The moderate volumes of Neogene clastic sediments imply overall low uplift and erosion rates until the present, favorable to morphological and lithological resistance effects in the landscapes. However, slope instabilities are not uncommon, locally leading to well-characterized processes, landforms, and deposits of gullying and mass wasting. Therefore, hazards related to slope processes should not be neglected in large parts of the study area.

Keywords Backwearing · Barreiras sediments · Downwearing · Fault scarp · Hazard · Landslide · Mass wasting · Neotectonics · Piedmont

J.-P. Peulvast and F. Bétard, *Landforms and Landscape Evolution of the Equatorial Margin of Northeast Brazil*, Springer Earth System Sciences,
DOI 10.1007/978-3-319-18203-2_5

5.1 Barreiras–Tibau Sediments and Landforms: A Geological Record of the Late Cenozoic Geomorphic Evolution

5.1.1 Fluvial to Coastal Sediments: Erosion and the Origin of the Late Cenozoic Sediments

The Barreiras sediments are part of a large clastic wedge of mainly Late Cenozoic age (Pessoa Neto 2003) that is found onshore (Barreiras s.s.) and offshore (Tibau Formation) along the margin of northeast Brazil. They overlie the weathered basement of the coastal piedmont, and also unconformably cover Late Cretaceous sediments in the Potiguar basin. Reduced to dissected remnants disconnected from source areas, these clastic deposits of Early Miocene age (see Sect. 2.5), and of limited thickness (10–80 m), slightly dip seaward, where they form the so-called "*tabuleiros.* " They belong to a piedmont system whose inner, eroded parts are exposed almost in the same plane as that of the "*tabuleiros.*" Their weathered substrate is clearly identifiable at the coast and along shallow valleys incised down to the basement. Generally flat, it is locally irregular, with low rounded hills and inselbergs protruding through the cover. Its buried distal parts as well as exposed proximal elements intersect each other at low angles and locally comprise slightly dissected surfaces. Both buried and exposed parts, the final shaping of which probably provided the Barreiras deposits, are diachronous in age. Such a piedmont geometry implies some kind of noncyclic evolution during the Neogene, such as described in distal parts of uplifting areas, in wide flexural zones where interplay between moderate uplift and variations of the base level only allowed a slow process of etch-planation and remained insufficient for triggering stages of deep river entrenchment until the final dissection (Klein 1993; Peulvast and Claudino Sales 2005).

According to most authors, the Barreiras deposits were produced by stripping of weathering profiles formed in humid conditions (Braga et al. 1981; Nascimento et al. 1981; Gunnell 1998) and were mainly deposited near the base level in wide alluvial plains as coalescent fluvial fans and braided systems, by rivers or debris flows, under semi-arid climates (e.g., Bigarella and Ab'Sáber 1964; Bigarella and Andrade 1964; Suguio et al. 1986). In classical views, they would be correlative of the last planation phases responding to the uplift of the Borborema region (Pd1 and P2 or Patos or Sertaneja: Bigarella and Ab'Sáber 1964; Bigarella and Andrade 1964; Bigarella 1975; Passos and Bigarella 1998; Jardim de Sá et al. 2005; Morais Neto et al. 2008). Recent sedimentological studies indicating tidal processes rather suggest deposition within a variety of mainly estuarine environments controlled by the combined effects of regional tectonics and relative sea-level rise related to the worldwide Oligo-Miocene and middle Miocene marine transgressions (Arai 2006; Rossetti et al. 2011, 2011).

In basement areas, remnants of waste deposits were mapped on parts of the Sertaneja surface, at some distance from the inner scarps, up to tens of kilometers. Deposits of argillaceous, often ferruginous, sands and gravels of alluvial to colluvial origins were mapped on slightly dissected pediplains at low altitudes, e.g., on proximal parts of the Jaguaribe pediplain and of the coastal piedmont, around the Baturité massif. Sparse deposits of coarse quartz and quartzite pebbles and gravels, several meters thick, are also found in proximal and even distal locations, on dissected fans and pediments (Fig. 5.1), and high fluvial terraces, in the same areas as well as south of the Potiguar basin. All of them reflect the occurrence of periods of active slope erosion and alluvial fan construction on proximal parts of the low surfaces. A slight entrenchment of valleys now places these deposits on interfluves, 20–30 m above the present valley floors and cuts them from their source areas, scarps, and inselberg slopes located up to 20 or 30 km away. This dissection, similar to that of the Barreiras sediments, may suggest a stratigraphic relationship between both sets of deposits.

The presence of coarse alluvial fans or terraces upstream of the main outcrops, in the vicinity of inner scarps and along several valleys indicates that slight scarp retreat, downwearing and even incipient dissection of the Sertaneja pediplain, were involved in this process. However, the limited thicknesses of the Barreiras sediments (less than 80 m, often less than 15 m) and their distribution on a narrow coastal strip (5–40 km) only reflect the denudation of a thin rock slice, since contributing areas are generally as wide as the buried ones or wider. Even taking into account the offshore part of the clastic wedge (Tibau Formation) does not change this statement. As measured on a synthetic profile in the Potiguar basin area, the mean thickness of this wedge (about 150 m at most on a 100 km wide strip) would represent at most 50 m of vertical erosion on a 300 km wide inland zone (i.e., the mean width of the

Fig. 5.1 Remnant of detrital deposits on the western piedmont of the Baturité massif. Rich in rounded quartz pebbles and gravels embedded in a sandy-clay matrix, this fanglomerate deposit was identified a tens of meters above surrounding present valley floors, 5–6 km from the western scarp of the Baturité massif. *Photograph* F. Bétard

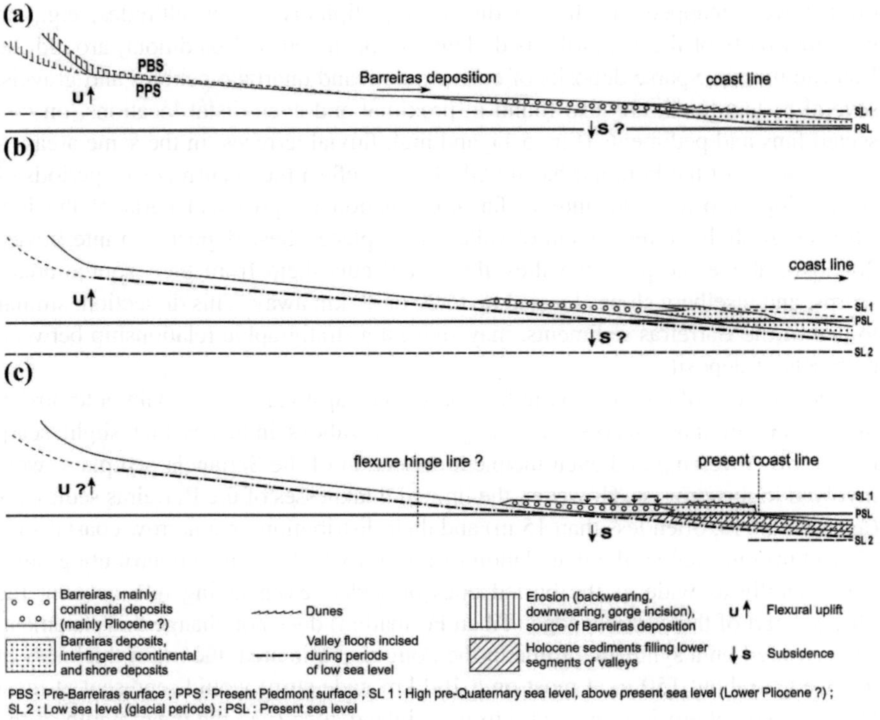

Fig. 5.2 Sketch diagram showing possible relationships between the Late Cenozoic tectonic evolution of the margin, the Pliocene and Quaternary sea levels and sequences of deposition and erosion of the Barreiras sediments. **a** Pliocene (?), high sea level, Barreiras deposition; **b** Period of low sea level stand (Pleistocene, glacial maximum), dissection; **c** Interglacial or present, infilling of low valleys. From Peulvast and Claudino Sales (2004)

Sertaneja surface). The eroded slice is probably thinner in the less uplifted coastal areas. Such moderate figures, also suggested by the identification of pre-Barreiras to pre-Cenomanian paleosurfaces on parts of the low surface, are comparable to average values measured from stratigraphy on the post-Cretaceous times (Peulvast et al. 2008). They do not imply any Neogene acceleration of erosion and uplift (Fig. 5.2). Later dissection of the low surfaces (Sertaneja surface), accompanied by scarp exaggeration and retreat, seems to correspond to a Late- and post-Miocene stage of vertical erosion observed in the whole Ceará region.

5.1.2 Sedimentation and Sea Level

Barreiras sedimentation has continued until the subsidence rate was superseded by rates of sea-level fall, before the initiation of the late Miocene, as suggested by (U-Th)/He dating of soils (Rossetti et al. 2013), with no record of subsequent transgressions at least until the late Quaternary. According to these authors, these datings conflict with the sea level recorded in other South American areas and also in other continents, suggesting that local factors such as intraplate tectonics might have played an important role in the regional geological history (formation of an additional accommodation space favoring the preservation of the Barreiras Formation). Thermochronological data mentioned higher indicate a crustal cooling event since 20 Myear ago in the eastern margin of Brazil. The onset of this cooling event is coeval with the deposition of the Barreiras Formation and it may indicate margin uplift and erosion during this period. The greater abundance of coarse continental deposits in the Barreiras layers around Fortaleza and to the west of the SPSZ compared to the eastern coastline may reflect the proximity of higher mountains and of a steeper zone of margin flexural uplift in the Baturité–Fortaleza region (Claudino Sales 2002). The gentle seaward dips of the Barreiras sediments may have been caused by the continuing effects of updoming of the Sertão hinterland.

According to the ages found for the deposition of the Barreiras sediments in northeast Brazil (between 22 and 16–17 Myear, i.e., before the middle Miocene highstand: Lima 2008), and for their weathering, until 7 Myear, there is indication of mainly warm and humid climatic conditions in the coastal area during this period. According to Rossetti et al. (2013), the prolonged sea-level lowstand recorded since the Tortonian may explain the prevalence of subaerial conditions and the lateritic soil formation from mid-Miocene up to late Quaternary.

5.2 Neotectonics and Landforms: Evidence and Uncertainties

5.2.1 Seismotectonic Activity, Stress Field and Neotectonics

The seismic activity in the study area has been investigated with the help of instrumental, historical, and paleoseismic data (Bezerra et al. 2011). A significant instrumental seismicity has been known in the study area since the 1960s and occurs within the first 1–12 km of the upper crust Ferreira et al. 1998). The heightened seismicity is characterized by moderate intensities (observed maximum: VII MM) and magnitudes mostly ≤5.0. It is mainly concentrated in a few areas close to the Senador Pompeu shear zone and related faults—specifically in of the Cascavel–Pacajus area of the coastal strip—(the Potiguar basin, its southern border, and the Aracati platform), and in the Fortaleza-Baturité area. A few earthquakes also occurred in the Sobral-Irauçuba area (Saadi and Torquato 1992; Ferreira et al. 1998) (Fig. 5.3). Other earthquakes of low magnitude ($m_b < 5.2$) are recorded in slightly inner regions, with epicenters located close to Precambrian shear zones (Jaguaribe and Portalegre fault zones) reactivated in the Cretaceous and later.

The historical seismicity is known since 1808, when an earthquake of estimated magnitude m_b = 4.0 and intensity MM > VI occurred near Açu. No coseismic surface rupture has hitherto been reported in northern Brazil (Assumpção 1992),

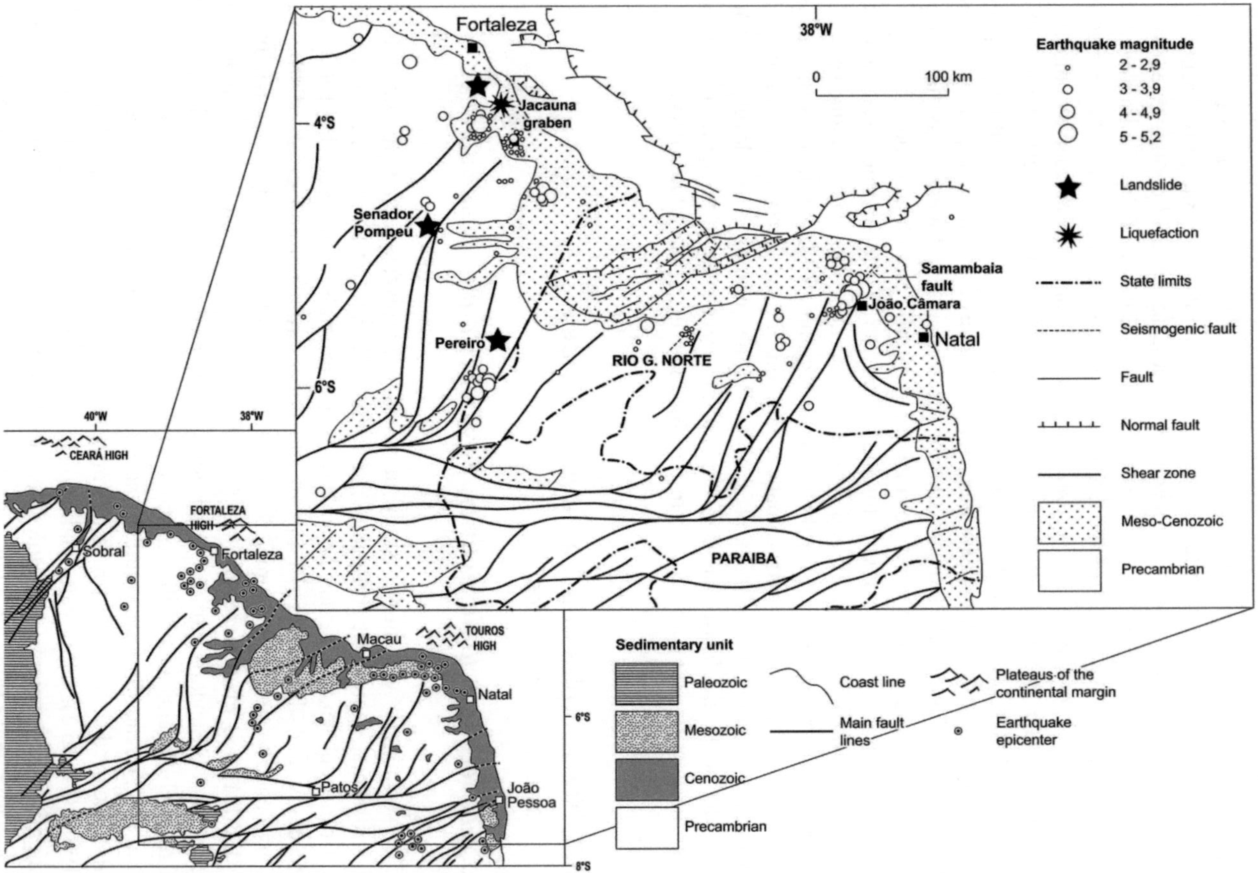

Fig. 5.3 Seismotectonic map of the study area. Compiled from Carneiro et al. (1989), Ferreira et al. (1998), and Bezerra et al. (2001), in Peulvast et al. (2006)

Fig. 5.4 Syn-sedimentary fault at a seacliff on the east coast of Ceará (Icapui, Redonda; see Fig. 5.1 for location): **a** normal fault affecting post-Turonian fluvial sandstone; **b** detail of **a**. Note that the footwall present soft-sediment deformation related to faulting. From Peulvast et al. (2006)

even by the most violent earthquake ever recorded in NE Brazil (Pacajus, November 20, 1980, 5.2 m_b) (Ferreira et al. 1998) or by the João Camara earthquake swarm in the period 1986–1989, in which an individual event reached 5.1 m_b (Takeya et al. 1989). Only a few phenomena of soil collapse and boulder falls from steep slopes were observed. The first soil collapse occurred after the Aracaticum earthquake swarm in March and April 1979, and the others occurred during the Pereiro earthquake of January 2, 1968, and the Senador Pompeu earthquake of February 23, 1968. This seismicity merely shows poor correlation with mapped faults.

Paleoseismic data are also available, mainly in coastal areas, from measurements on late Cenozoic, mainly Holocene, deposits. Along the Carnaubais fault, the base of the Pliocene sequence is downfaulted by about 60 m southwest of São Bento (Bezerra and Vita-Finzi 2000), suggesting that the minimum average slip rate might be 0.01 mm year^{-1}. Seismic-induced liquefaction features have been found in the study area. They include dikes, pillars and pockets, which occur in once-water-saturated Quaternary alluvial gravels in the Açu and Jaguaribe River valleys (Bezerra et al. 2005). These features are evidence of earthquakes of 5.5 M_L and 5.6 M_S or greater, which would have occurred on the region as recently as 4860–4570 year BP (Bezerra et al. 2001). They are also visible in cliffs and shore platforms of the east coast of Ceará (Fig. 5.4), especially in those of the Ponta-Grossa-Peroba area (Icapui), where they have been interpreted as neotectonic since the deformed sediments are considered as belonging to the Barreiras Formation (Sousa et al. 2008a, b; Fig. 5.5a). The liquefaction structures represent secondary features associated with the reactivation of strike-slip fault zones (Bezerra et al. 2005).

Analysis of focal mechanism indicates EW- to NW–SE-oriented maximum compression, parallel to the coastline, and NS to NE–SW-oriented extension (Ferreira et al. 1998). Since the Pliocene, these orientations roughly correspond to the direction of movement of the South American plate, which meets strong resistance on its western border (Andes), and to the flexural movement located in the coastal area (Assumpção 1992). In the study area, this activity is in keeping with

Fig. 5.5 Deformed post-Turonian sediments and unconformable deposits (allochtonous laterite, paleodunes). **a** Ponta Peroba (Icapui): Faulted Barreiras sandstones and conglomerates unconformably overlain by allochtonous laterite and post-Barreiras aeolian sediments. **b** Camocim: folded Barreiras conglomerates, with undulated topography buried by post-Barreiras aeolian sediments. *Photographs* J.P. Peulvast

several tectonic pulses that have caused the development of strike-slip and normal faulting in coastal areas since the Cretaceous.

Active faults are suspected to control a well-developed horst and graben morphology and the drainage, alluvial, and coastal patterns in eastern Rio Grande do Norte between Natal and João Pessoa (Bezerra et al. 2001). However, the only scarp related to an active fault in the study area was recognized on the coast of the Potiguar basin near São Bento (Fig. 5.3). It forms cliffs up to 7 m high southeast of the Carnaubais fault, in a marine bioclastic grainstone that interfingers with the Barreiras Formation. The base of this sequence is downfaulted by about 60 m southwest of São Bento (Bezerra and Vita-Finzi 2000). Other indications of possible neotectonic movements were obtained from the analysis of the geometry of

Tertiary and Quaternary fluvial and marine depocenters in the Açu River delta near Macau, where NW- and NE-trending normal fault movements cut across marine and fluvial sediments dated to 30,190 ± 370 year BP with offsets of 30 m (Silva 1991). Neotectonic movements are also inferred from the identification in the Barreiras Formation sedimentary rocks of small normal faults trending N40°–60° and from clastic dikes NW of the Potiguar basin (Fazenda Retirinho) (Torquato et al. 1991) and in NW Ceará (Camocim) (Saadi and Torquato 1992).

In most cases, active faults recognized in the study area do not present topographic expression, which should be implied in small throws and/or erosion rates higher than tectonic rates (Bezerra et al. 2011). Moreover, they mostly occur along the continental margin, in sediments of the Barreiras Formation and alluvial deposits whose weak resistance to erosion is not favorable to preservation of tectonic landforms. The study area lacks equivalents of the grabens of the eastern coast (Rio Grande do Norte and Paraíba), along which river erosion has allowed a moderate topographic expression of the most outstanding faults (Bezerra et al. 2001, 2011). However, linear (fluvial) erosion along some fault structures of the Macau region, mainly during the Pleistocene periods of low sea levels, may have exaggerated the depth of the downfaulted compartments and contributed to controlling further sedimentation (Fig. 5.6). At last, the main reason for the absence of recognized neotectonic landforms probably lies in the focal mechanisms in northeastern Brazil, indicating that the current faulting regime is strike slip (Bezerra and Vita-Finzi 2000; Bezerra et al. 2011). Outside transpressive or transtensional zones (restraining or releasing bends), this does not favor high scarps or significant relief expression of neotectonic movements.

To the south of the Potiguar basin, recent uplift was postulated to explain how the fluvial sediments at Martins may cap high mesas 600–700 m above the low-lying Lajes depression and Chapada do Apodi (Jardim de Sá et al. 1999). By analyzing drainage orientations around the Serra de Santana, Menezes et al. (1997) and Barros et al. (1997) looked for evidence of recent deformations in and around this plateau. Few drainage anomalies were found, suggesting that this mesa belongs to a wider tectonic compartment in which no neotectonic activity is reported. Indeed, according to some authors, this altitude might result from uplift and slight southward tilting along the so-called Lajes lineament at the southern border of the Potiguar basin. Such a structure would reflect regional flexure caused by sedimentary overload on the continental margin (Jardim de Sá et al. 1999). Drainage orientations and possible anomalies were also studied with the purpose of detecting neotectonic activity in SE Ceará and W Paraiba, around Chapada do Araripe, without any significant result (Fonseca and Melo 1997).

5.2.2 The Coastal Strip: Neotectonic Control on Landforms?

Outside the most conspicuous neotectonic features described in the Potiguar basin (antiforms of the Serra do Mel and Serra de Mossoró: Maia and Bezerra 2014a, b),

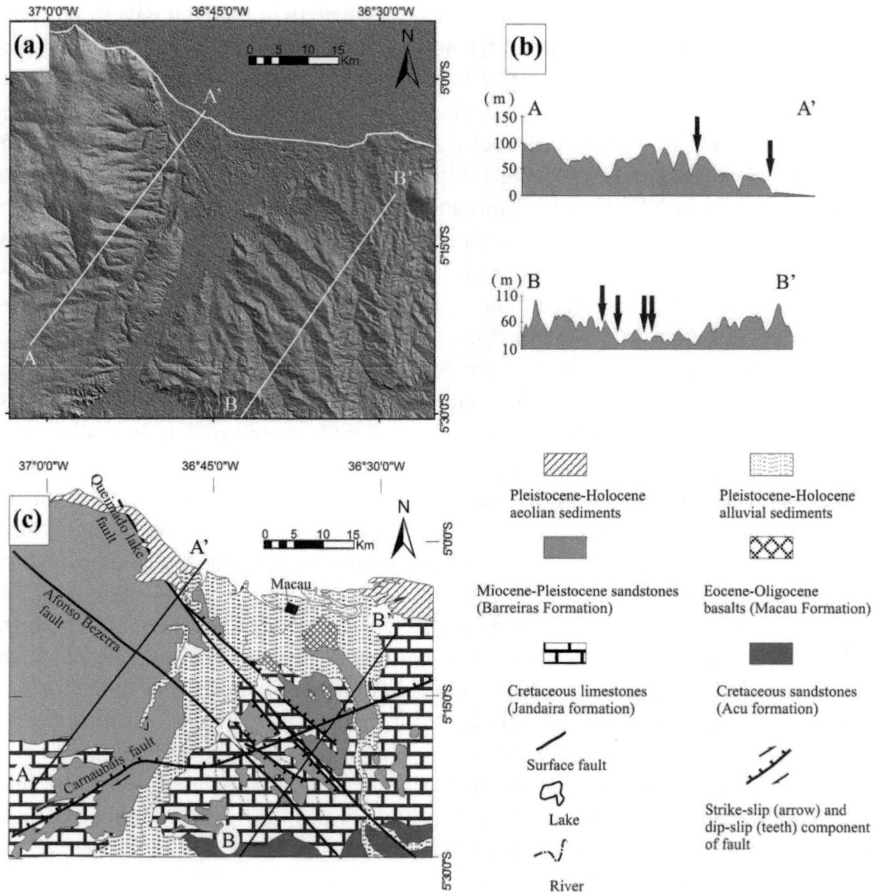

Fig. 5.6 Topographic and geologic signatures of strike-slip faults in the central part of the Potiguar basin: **a** SRTM topography illuminated from NE; **b** topographic cross-sections; *arrows* indicate trace of faults on surface; **c** simplified geologic map. From Peulvast et al. (2006)

the major faults along which a Late Cenozoic activity was recognized do not show significant topographic expression (Peulvast et al. 2006). The best example of young but presently inactive structures is that of the Icapui–Ponta Grossa area, in the onshore Potiguar basin (Sousa et al. 2008a, b). Here, the Barreiras sediments exposed in the seacliff display a complex array of faults and folds corresponding to extensional as well as contractional oblique structures related to a transcurrent system bearing NE trends. Continuing older phases of deformation of Neocretaceous to Paleogene age, at least part of this tectonic phase occurred simultaneously with lateritic soil formation, as suggested by the ductile deformation (folds) of presently indurated deposits (Fig. 5.5b), i.e. before 7 Myear. Later on, these structures were completely eroded and then unconformably covered by the post-Barreiras sands of the Potengi Formation.

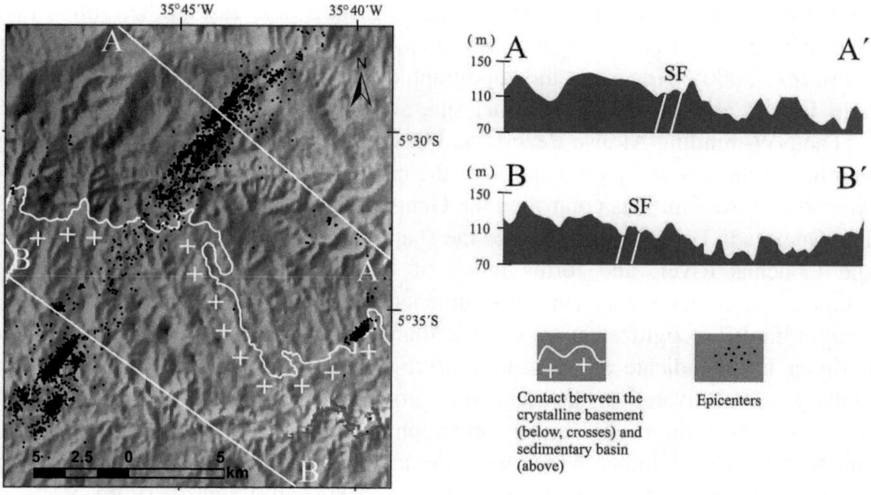

Fig. 5.7 The Samambaia fault and related topography derived from SRTM DEM. Note in cross-sections A–A′ and B–B′ that crustal blocks across the fault do not show significant topographic changes. From Peulvast et al. (2006)

Many of the scarps described along faults recognized as active features in the Potiguar basin are too small, less than 5 m high, to be identified on regional topographic maps. However, along the coast, several fault-bounded valleys are characterized by topographic breaks 20–40 m high between Cretaceous and late Tertiary rocks and Quaternary sediments. These fault scarps are faint and degraded and have been attacked by weathering and erosional processes. Colluvium has accumulated, and debris slopes have grown due to fault scarp retreat. Free faces are uncommon or absent because most fault scarps are capped by debris, vegetation, and soil, indicating that most have not been active recently, despite their late Tertiary to Quaternary age. The intervals at which most of these faults offset the surface seem to be similar or slower than the rates at which surface processes modify the landscape, which are of the order of 0.001 mm year^{-1} (Peulvast et al. 2006; Bezerra et al. 2011).

One of the best examples of seismogenic faulting in Brazil corresponds to the Samambaia fault, and its related topographic features (Takeya et al. 1989; Bezerra et al. 2011). No topographic contrast occurs along this fault (Fig. 5.7).

Marks of seismic activity are also recognized along the NE continuation of the Portalegre shear zone, called Carnaubais fault zone in the Potiguar basin (Fig. 5.3). The Carnaubais fault zone deforms both the Açu and Jandaira formations (Bertani et al. 1990) and seems to have been reactivated from the late Tertiary to the Quaternary, since it has controlled the sedimentation during this period (De Castro et al. 2012). Along the coast, east of Macau, it forms the boundary between the Barreiras Formation and Quaternary coastal sediments composed mainly of sand dunes and tidal flats. Inland, the lithological boundary between the Jandaira and the Barreiras formations to the SW of São Bento is roughly parallel to the Carnaubais fault. Coseismic uplift is reported to have occurred to the east of the fault, where rapid emergence of at least 4 m occurred ca 4080–2790 cal. year BP (Bezerra and Vita-Finzi 2000). However, the topographic contrast between these lithological units is only about 5 m in the former site, and about 15 m in the latter).

The NW-trending Afonso Bezerra and Queimado Lake faults control Quaternary sedimentation and drainage patterns in the central part of the Potiguar Basin. The Afonso Bezerra fault has controlled the Gangora River and Vargem de Cima Lake. The Queimado Lake fault influences the shapes of Queimado Lake, Cavalos River and Conchas River, and forms much of the boundary between the Barreiras Formation and Quaternary coastal sediments. Kinematic indicators obtained along straight faults recognized across more than 100 km in the central part of the Potiguar basin indicate an important strike-slip component of movement. These faults were reactivated in the Cenozoic, producing 30–35 m high degraded fault scarps. On the contrary, no similar topographic expression is known along the EW- and NNW-trending faults, which were the locus of seismic activity since 1980 on the western border of the Potiguar basin (Fig. 5.3) (Ferreira et al. 1998).

5.2.3 Inland Scarps: Weak Evidence of Neotectonics

Inland, most faults along or near which recent seismotectonic activity is recorded control important features of the regional relief, at least locally. Many of these features are oblique or perpendicular to the coastline. They correspond to Precambrian shear zones already reactivated during the Cretaceous rifting: Transbrasiliano, Senador Pompeu, Jaguaribe, and Portalegre shear zones. Some of these shear zones (Sobral-Pedro II and Senador Pompeu) have played major roles in decoupling between crustal compartments of the passive margin that have undergone different post-breakup tectonic evolutions (Matos 1992; De Castro et al. 1997).

According to the distribution of seismic zones, the main scarps and other landforms whose existence or shapes might be partly or completely explained by neotectonics are the following:

- *the discontinuous basement scarp, topped by sandstone tables, that forms the northern edge of the aligned spurs and buttes of Pereiro, Portalegre, Martins, João do Vale, and Santana* and overlooks the Chapada do Apodi, beyond the wide Apodi-Lages depression underlain by the exhumed basement. It corresponds to a residual fault scarp which has retreated from remote rift structures (Carnaubais and Apodi faults) (Fig. 5.8a), mainly before the deposition of the post-rift cover that they slightly deform owing to local fault reactivation within the Potiguar basin during the Cenozoic. The pre-Cenomanian bevelling of the rift structures strongly suggests the old age and the erosional origin of this scarp. Its origin is also indicated by the lack of any important parallel fault or flexure zone between these rift structures and the northernmost edge of the Borborema uplands (Portalegre, Martins, João do Vale, and Santana plateaus).

(a)
(b)
(c)
(d)

◀ **Fig. 5.8** Inland scarps and tectonics related to faulting. **a** The north edge of the Serra do Martins, a residual fault scarp, as seen to the east. Knick, marble buttresses (*foreground*), inselbergs at the contact with the Sertaneja surface (which derives from the exhumed pre-Cenomanian surface). In the distance, to the ESE: the Serra João do Vale, another remnant of the south shoulder of the Potiguar rift. Located 30–40 km to the north (i.e. to the *left*) in the Potiguar basin, the master faults are not visible in the area represented here since the pediment only appears over 5–10 km wide in the picture. **b** The Pereiro fault scarp, seen to SE from the BR 116 road between Icó and Jaguaribe. Note the weak sinuosity of the escarpment, which coincides with the outer limit of a syn-orogenic granitic intrusion. Presence of short steep and hanging valleys, with small alluvial fans. Gneissic pediment in the foreground. **c** The east escarpment of the Serra da Aratanha (south of Fortaleza), as seen to the south. Its straight outlines and profile look like those of a fresh looking fault scarp, but as in the eastern border of the Baturité massif, this scarp must be considered as erosional. **d** The SE fault or fault line scarp of the Serra da Meruoca. The slope angle of the facets (25–30°) is slightly weaker than on active fault line scarps, whereas the low sinuosity and the location of the scarp on the fault line may be explained by strong lithological control as well as by neotectonics. The presence of a short fault line corridor between the conglomerate ridge visible in front of the main scarp (*middle*) suggests a decisive role of differential erosion, at least in the formation of the basal part of the scarp. *Photographs* J.P. Peulvast

Moreover, the historic seismic activity is concentrated around fault zones that obliquely intersect the so-called Lajes lineament, mainly the Portalegre shear zone (Bezerra and Vita-Finzi 2000), and other fault zones located to the east (Fig. 5.3). Although important at regional scale and in spite of the Cretaceous reactivation of at least one of them (Portalegre shear zone) (Nóbrega et al. 2005), these faults do not control important morphological features at the scarp or below, except for the water gap through which the Piranhas River crosses the EW alignment of buttes.

- *The Baturité-Aratanha scarp and the Senador Pompeu shear zone*. The 500–600 m high escarpment that forms the eastern edge of the Baturité and Aratanha massifs, south of Fortaleza, is devoid of clear structural control. Most peaks, culminating ridges, and outstanding escarpments are shaped into the folded quartzite layers which form the skeletal structure of the massif. No fault line is identified along the sinuous scarp and its multiple embayments, located 50 km NW from the NE–SW Senador Pompeu shear zone and from the Choró River. The dissected pediments that form the floor of the embayments and merge with the slightly undulating Sertaneja pediplain to the southeast do not display any topographic break that might indicate significant movements close to the massif. Even along the straight eastern scarp of the Aratanha massif, to the north, we consider the outlines as erosional (Fig. 5.8b). Records of the historic seismic activity in this area (Saadi and Torquato 1992) suggest a possible influence of moderate neotectonic movements along minor faults, but no mark of possible active tectonics was identified (Peulvast et al. 2006).
- *the western edge of the granitic Pereiro massif.* All along the scarp, the coincidence of his outline with the intrusive contact clearly indicates a contribution of differential erosion associated to the formation of the lower surface (Fig. 5.8c). Macroseismic data indicate that the seismic activity is concentrated in the west part of the granitic massif (Ferreira and Assumpção 1983). It may be related to fault segments in the plateau area to the NE of Icó, where the

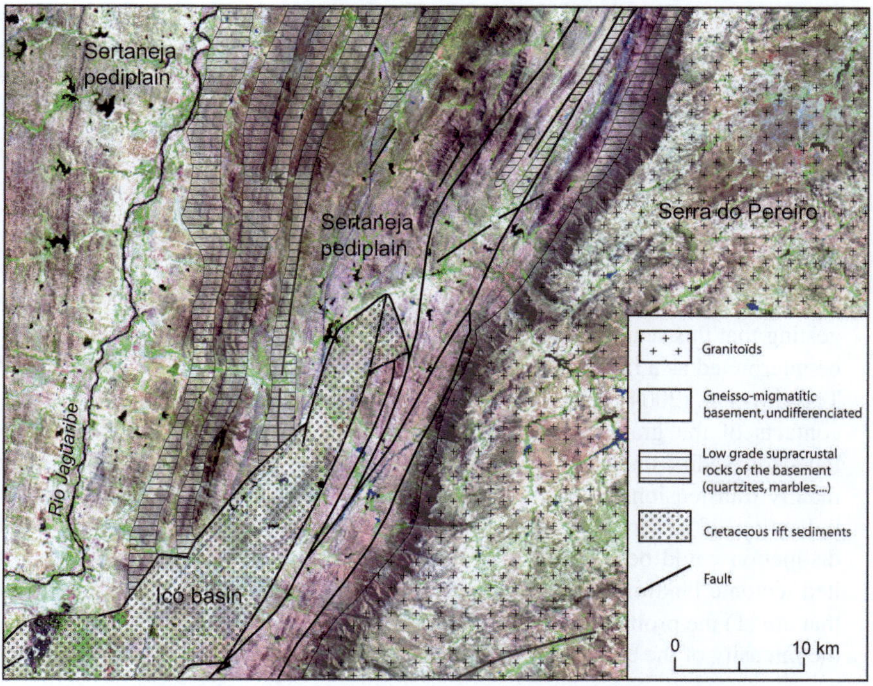

Fig. 5.9 Geological map (CPRM 2003) superposed on a Landsat picture of the Pereiro fault scarp. Note the coincidence between the escarpment, a major fault, and the contact between granite and gneiss, that may account for tectonic as well as lithological (differential erosion) origin. From Peulvast et al. (2006)

escarpment splits into minor crests on both sides of the fault zone (Fig. 5.9). Here, and to the north, it appears as a residual fault scarp controlled by the intrusive contact parallel to the fault zone, but having undergone some erosional retreat in relation to the development of a narrow strip (1–5 km) of rugged pediments and alluvial fans. A recent study of this massif points out the possibility of differential tectonic uplift along reactivated Brasiliano shear zones (Gurgel et al. 2013). However, slip rates measured along the most active faults of northeast Brazil, which are located in coastal areas (0.0075–0.01 mm year^{-1} since the Miocene), amount to 180 m of vertical displacement in Cretaceous units in the Paraíba basin (Bezerra et al. 2001; Nogueira et al. 2010). Such values suggest rather minor and localized neotectonic contributions to the formation of ridges and mountains in the landscape at some distance from the most active seismic zones. Even the presence of lateritic remnants on the Pereiro plateau, possibly as old as 20 Myear (Lima 2008), does not clearly support evidence of strong local uplift, given that laterite can form in various topographic and altitudinal contexts (Rossetti 2004). Moreover, a neotectonic origin would not be compatible with the presence of the exhumed and undeformed palaeopiedmont of Cretaceous age at the northern tip of the Serra do Pereiro

(Peulvast and Claudino Sales 2004). This feature, therefore, rather suggests long-term preservation of scarps (inherited fault scarps, some of them possibly buried and later exhumed), as well as height enhancement of fault line scarps and other lithologically controlled scarps (Peulvast et al. 2006; Peulvast and Bétard 2013) during a long-term process of downwearing and differential erosion in response to regional uplift.

- *The Meruoca scarp*. The highest escarpment (SE wall) of this high granitic massif coincides with the Transbrasiliano dextral shear zone. In spite of the local presence of perched valleys and triangular facets similar to those of fresh fault scarps, outstanding erosive remnants are found on the opposite block, suggesting that this scarp mainly displays differential erosion features and thus may be interpreted as a fault line scarp (Peulvast et al. 2006; Fig. 5.8d). According to Teixeira et al. (2004), only subhorizontal striae were found on fault planes along contacts of the granite with the sediments. In addition, the downfaulting of Paleozoic sediments along this shear zone near Santana do Acarau appears to be mainly transtensional (Destro et al. 1994; Galvão and da Silva 2002), without indication of generalized post-Paleozoic vertical movements. Here again, no distinction could be established between features of differential erosion, inherited tectonic landforms, and possible results of recent faulting. The reasons for that are (1) the probability of a prolonged continuity in tectonic regimes, and (2) the intensity of the erosional shaping after exhumation from the Paleozoic cover.

The first of these landforms is the only part of the discontinuous marginal escarpment where an intervention of neotectonics has been proposed (Barros et al. 1997; Jardim de Sá et al. 1999). No local structural feature explains its position (Fig. 3.2). The same problem arises for many parts of the eastern scarp of the Baturité-Aratanha massif. On the contrary, in the last two examples, the coincidence with intrusive contacts suggests a contribution of differential erosion associated to the preservation of residual mountains above the low erosion surface (Peulvast and Bétard 2013).

Finally, only small-scale features related to neotectonic activity in late Cenozoic deposits have been recognized in this context of passive margin, where tectonic activity remains moderate (Bezerra and Vita-Finzi 2000). At larger scales, only effects of long-lasting movements and long-term evolutions may be accurately measured, on the basis of morphostratigraphic arguments.

5.2.4 Asymmetry of Drainage Basins: The Middle Jaguaribe Basin

The drainage pattern of the study area strongly reflects the Cretaceous tectonic heritage, since the main rivers either follow well-identified Neocomian rift structures or flow toward them from their shoulders, as in Leeder and Gawthorpe's models (1987). For example, it may explain the strong asymmetry of the middle

Jaguaribe basin, where the main river trunk, downstream from Orós, receives only a short tributary (Figueiredo River) on its right (eastern) side, whereas on the opposite side it receives long rivers draining the northwest shoulder of the Potiguar rift, reflecting a former hanging/foot wall slope opposition related to the Jaguaribe fault system (see Fig. 3.15). Thus, downstream of the confluence with Salgado River, the Asymmetry Factor (Keller and Pinter 1996) of the watershed is (27,400 km^2: 8130 km^2) 0.100 = 337.

However, this pattern may also reflect an eastward attraction related to the post-rift and possibly recent subsidence of the Potiguar basin. This is suggested by the trend of the Banabuiú River, which turns east and leaves the shallow corridor excavated along the Senador Pompeu shear zone. It crosses the Orós-Serra do Felix ridge rather than continuing NE toward the Piranji River—this river being the only important one in Ceará that is not fed by sources located in the highlands. A former Banabuiú-Piranji River could have been captured by tributaries of the lower Jaguaribe River, presently the lower Banabuiú and Sitia rivers, more easily deepened owing to its regional dip toward the Potiguar basin and the weak resistance of its sediments.

5.2.5 Poorly Characterized Neotectonic Movements in the Interior of Northeast Brazil

Precise stratigraphic indications on the amplitude and rates of tectonic movements were only obtained SE of the Senador Pompeu shear zone and in south Ceará. In spite of their potential interest, observations on neotectonic movements gave more limited results. All of them were obtained in coastal areas, from measurements on Holocene, Pleistocene, and late Cenozoic deposits. Along the Carnaubais fault, the minimum average slip rate might be 0.01 mm $year^{-1}$. Higher rates may be locally recorded in Pleistocene marine deposits, with fault throws as high as 10–12 m, as shown by the uplift of the youngest Pleistocene marine terrace around Touros (oxygene-isotope substage 5e, 117–110 ka BP) (Barretto et al. 2002), yielding rates up to 0.4 mm $year^{-1}$. Inland, slip rates are probably much lower. Recurrence intervals of earthquakes responsible for coseismic deformations are not known since they are longer than historical settlements (Bezerra et al. 2011).

The generally low post-rift deformation rates observed in most cases are comparable with long-term uplift and denudation rates obtained by morphostratigraphic methods in the Jaguaribe–Piranhas hemicycle, at most 10 mm ka^{-1} in southern Ceará, and much less in the weakly uplifted coastal area of the Jaguaribe–Potiguar compartment (Peulvast et al. 2008). Cenozoic tectonics and magmatism related to the evolution of the adjoining oceanic area have been considered as possible controls in the morphostructural evolution of offshore and onshore domains. The strong increase in the proportion of clastic sedimentation recorded in the offshore Potiguar basin since the Miocene is interpreted as a result of such a Cenozoic uplift (Pessoa Neto 2003). Actually, it does not reflect a strongly increased erosion rate, since the mean thickness of the Late Miocene and younger clastic Barreiras and

Tibau sediments (about 150 m at most on a 100 km wide strip, measured on profiles of the Potiguar basin) would represent at most 50 m of vertical erosion on a 300 km wide inland zone, i.e., a rate <10 mm ka^{-1}, similar to the average rate obtained for the post-Cenomanian times (see Sect. 4.3.1). In such conditions, no accelerated uplift stage is recorded regionally.

Although local movements are possible, no significant escarpment is expected to have been formed by neotectonic activity, unless exaggerated by shore or fluvial erosion in coastal areas (Bezerra et al. 2001, 2011). Together with the limited magnitude of this tectonic activity, long recurrence intervals between seismotectonic events probably explain why the morphological expression of faults is often faint or ambiguous in such a context of low slip rates. Moreover, strike-slip components of movement prevail in the whole area. Therefore, the occurrence of expressive tectonic landforms related to neotectonic movements remains unlikely on this passive margin, where possible confusions between these and differential erosion landforms of various ages obscure their identification.

5.3 Late Cenozoic and Present-Day Erosion Dynamics of Scarps and Piedmonts

The steep slopes forming the marginal scarp of the semi-arid northern Brazilian "Nordeste" and bounding inner residual plateaus and ridges rising up to 800–1200 m overlook a low composite surface (Sertaneja surface) merging seaward with a well-developed coastal erosional-gradational piedmont. Recent or active mass wasting landforms, dissected pediments and sparse debris fans visible at the base of some escarpments suggest recent and current erosive activity and possible slope retreat, although most of these forms are decoupled from major valleys. Therefore, the question arises of the age, origin, evolution and stability (or instability) of these scarps, and of the signification of the observed or inferred dynamics in relation with long-term trends of evolution. Beyond the indications obtained on types of slope dynamics and on the factors that control their occurrence, this can throw light on duration and more general factors that can explain slope retreat (backwearing) and/or piedmont downwearing outside the main drainage axes, i.e., a fundamental issue in the understanding of planation processes and long-term landform evolution in tropical areas (Peulvast and Bétard 2013).

5.3.1 Structural Controls on Late Cenozoic Scarp Evolution and Hillslope Dynamics

The long-term morphostructural context is that of shallow basin inversion that has occurred in the Ceará area of NE Brazil since Cenomanian times. Uplift resulted in the topographic inversion of the post-rift basins, exhumation of buried surfaces/

stratigraphic unconformities, dissection of the residual Cariri–Potiguar footwall uplands, and expansion of the erosional Sertaneja and coastal plains (see Chap. 4). The types of structural controls on scarp evolution and hillslope dynamics differ according to the nature of the bedrock: crystalline or sedimentary. Among the basement scarps, two main types were analyzed. In the first one (Pereiro and Meruoca scarps: Fig. 5.8c, d), a coincidence with intrusive contacts, highlighting differential erosion features, suggests limited possibilities of scarp retreat. Both granitic scarps are disconnected from the main valleys excavated along master faults (Jaguaribe, Acarau) by 5–15 km wide pediments. In the second case (south Potiguar and Baturité-Aratanha scarps), important scarp retreat by dissection and dismantling of old Cretaceous rift shoulders or large flexure hinges may have occurred during the Late Cenozoic together with the formation of erosion piedmonts, possibly helped by the marginal flexuration that reduced the thickness of the rock slices to be removed in the vicinity of the coast and of the basins. However, the high sinuosity of these scarps ($S > 2$) seems to correspond to the complexity of local structural control that fixed scarp outlines in a long-lasting evolution: such controls are best explained in a context of evolution by etch-planation and downwearing limited to the most easily weathered rock volumes.

In sedimentary scarps such as those of the Chapada do Araripe and Serra da Ibiapaba, the outlines, profiles and possibilities of retreat are mainly controlled by the respective thicknesses of resistant and soft (and/or weatherable) rocks, and by local dips, especially in the cap layer (Peulvast and Vanney 2001; Peulvast and Bétard 2015). Weak dips are more favorable to active retreat, at least locally, in the form of sapping and development of box canyons such as those that indent the rim of the Chapada do Araripe to the north (Santana do Cariri), the east (Jardim) and the south (Ipubi). This pattern allows the dissection of the plateaus in elongated promontories and outliers as well as the shaping of wide orthoclinal depressions. On the contrary, higher dips impede or slow down scarp retreat, and the outlines remain weakly indented, except along cataclinal valleys. Observed in the Serra da Ibiapaba, this situation also prevents the preservation of outliers. As shown in the Araripe scarp (Peulvast and Bétard 2015), these patterns also control the possibilities of deep-seated mass wasting.

5.3.2 Structural and Other Controls on Present-Day Morphodynamics

5.3.2.1 Basement Scarps

Whereas most granitic and migmatitic escarpments display steep and bare walls, frequently shaped into rounded domes or sharp needles (Pereiro, north of the Baturité massif; Figs. 5.8c and 5.10a), or steep facets with straight profiles (Aratanha, Meruoca SE; Fig. 5.8b, d), the wall morphology is more irregular in the sinuous scarps shaped into metamorphic rocks (Baturité E; see Fig. 3.11). High rock walls are

Fig. 5.10 Scarp morphology in basement rocks. **a** The northern scarp of the Baturité massif, near Palmacia. Migmatite ridges and domes (sugarloaves) shaped by cooperation of deep weathering and saprolite stripping into the most dissected part of the massif. No identified fault line between the mountain and the erosion plain visible in the foreground. **b** Quartzite scarp (cuesta) above the Capistrano embayment, eastern rim of the Baturité massif. Mass wasting and rockfall deposits. **c** Mass wasting and debris flow deposits in the eastern edge of the granitic Meruoca massif near Massapê. Recent and active saprolite stripping, probably enhanced by human activity (deforestation). *Photographs* J.P. Peulvast

visible in the western and southern slopes of this massif, where they are cut into thick quartzite layers dipping opposite to the slope inclination (Pico Alto, Capistrano; Fig. 5.10b). However, minor, non-coalescent alluvial fans are found at the base of both types of scarps, below short and steep valleys or gullies. In several locations, mainly in the driest and least vegetated parts of granitic massifs (northern Meruoca walls; Fig. 5.10c), but also on humid, forested slopes (Serra de Maranguape), we found fresh forms and deposits of debris flows, probably formed after heavy rains of recent active rain seasons (2003, for instance). Their sparse distribution reflects that of fracture zones or benches where significant volumes of saprolite (mainly grus and blocks) were temporarily retained in otherwise steep walls before being removed. The same was observed below many quartzitic ledges, where waste volumes may locally be important (western margin of the Baturité massif, west of Aratuba).

Active or inherited deep weathering is the main factor to explain superficial mass movements observed in hillslopes and scarps of the humid Baturité massif and, more generally, of parts of basement landforms where thick saprolites are formed and/or preserved (Bétard 2007). Such movements are mainly limited to the flanks of convex hills cut into the ferrallitic saprolites of the summit plateau and, more exceptionally to steeper slopes of the deeply dissected part of the massif, to the north; they do not seem to be extensively involved in scarp retreat processes at the periphery (Bastos 2012).

On the contrary, slope evolution is slower in places where lithology and bioclimatic conditions are less prone to deep weathering, allowing the production of minor waste quantities, often containing large blocks prepared by decompression and other processes in rock walls. Only non-coalescent alluvial fans are found at the base of scarps, below short and steep valleys or gullies. In the driest and least vegetated parts of granitic massifs, the sparse distribution of fresh forms and deposits of debris flows reflects that of fracture zones or benches where significant volumes of fine-grained saprolite can have been temporarily retained in otherwise steep walls before being removed. The same was observed below many quartzitic ledges, where waste volumes may locally be important.

Only a few gullies and deposits are connected to the drainage system of the surrounding erosion surfaces, since the entrenchment of lower river segments is observed at distance from the scarps. Most alluvial fans are built or preserved on proximal, nondissected remnants of pediments. However, some mountain valleys are continued through dissected remnants of pediments, mainly along sinuous scarps, even far from the coast (80–120 km: Baturité, Serra do Martins, Ibiapaba glint), showing that slight vertical incision could occur in some proximal parts of the piedmonts, after late pedimentation stages. However, in the Ubajara and de Viçosa do Ceará regions, where the sandstone ledge of the Ibiapaba glint is particularly high above the exhumed basement, no significant retreat linked to this vertical incision could be observed, except in the main sapping alcoves.

5.3.2.2 Tableland Scarps: Chapada do Araripe and Serra da Ibiapaba

The morphology of the north and NE rim of the Chapada do Araripe above the Cariri depression displays the most spectacular marks of a fast and active evolution, which seems to be exceptional in this part of the semi-arid Brazilian "Nordeste" (Peulvast et al. 2011; Peulvast and Bétard 2015). Over a length of 25 km, the escarpment forms two large adjacent reentrants (Crato, Barbalha), more or less semicircular and 10 km wide each, which open to NE. The 250–300 m high wall overlooks short concave slopes, either regular or cut by narrow benches or hummocky topographies, which connect its base (750 m a.s.l.) with wide systems of dissected glacis gently sloping (3.5 %) toward the center of the depressions, from 600 to 460 m a.s.l., over 2–5 km.

The Crato and Barbalha reentrants, where the scarp is the highest, display the most spectacular and diversified landforms and deposits related to gravitational dynamics. Simple or complex landslide landforms and deposits are found along concave wall segments. In the best characterized landslides, prominent deposits form thick hilly lobes over pediment roots. Slumps are the simplest of them. They are often prolonged by flat lobes of waste or long debris flows. Large rotated and

Fig. 5.11 Mass movement landforms and deposits in the sandstone scarp of the Chapada do Araripe. **a** The Meio rotational landslide (Barbalha), as seen by Google Earth. The photo shows a section in a huge block of Exu sandstone tilted to SW. **b** Stepped tilted blocks SSE of Barbalha (Flores): a km long slide scar splits into various benches reflecting a large (pluri-hectometric) retreat of the plateau rim. **c** Debris flow sediments: the lower Salamanca debris flow, upstream of Cabaceiras. **d** The head of the Carretão landslide, south of Crato, seen to NW: large scale slump, below a 1.5 km long and 180 m high scar with smooth concave outline and profile. The angle of slope decreases downward from 30 to 25–20°. *Below*, a chaotic topography with 60–80 m high hills incised by four deep and narrow valleys disconnected from the scar extends over a width of 2 km. In the *background* spur and funnel topography, at the head of large debris flows. *Photographs* J.P. Peulvast

tilted blocks are identified in some of them (Meio; Fig. 5.11a). Along most other concave segments of the scarp, coalescent systems of wide (1–2 km) and short (generally <1 km) backward tilted blocks or debris lobes were observed, suggesting that large and thick pieces of the sandstone cap were destabilized *en bloc* from the plateau edge along curved failure surfaces, and dislocated during their movement. To the SE of Barbalha, a kilometer long slide scar splits into various benches reflecting the large (hectometric) retreat of the plateau rim associated with the most important of these landslides (Fig. 5.11b). In all cases, movements occurred along mechanically weak levels of the sedimentary series (clays, marls and gypsum). Possibly initiated by shearing and horizontal translation along these levels, they are combined with rotational sliding along listric fault planes developed in the thick overlying sandstone layer.

Below systems of semicircular funnels or alcoves carved into other parts of the walls in the main amphitheaters, or downstream of slumps and debris tongues, the floors of some of the main valleys are partly occupied by elongated tongues of block-rich material emplaced by large debris flows (Fig. 5.11c). In the alcoves, the sandstone walls only bear scattered marks of recent block falls or rockslides. Minor accumulations of waste—weathered sandstone blocks in abundant matrix of reddish clayey sand—are found below, and spring water often flows out from the discontinuous talus. Therefore, no significant reconstitution of a colluvium reserve seems to have occurred after the stripping events at the origin of the debris flows. Measured lengths vary from 2.5 to 6.5 km, for vertical ranges of 250–350 m. Their widths are generally between 150 and 250 m but debris flow deposits form a 800–1000 m wide lobe between the Riacho Saco twin valleys. Using observed thicknesses, minimum volumes of 2.5 Mm^3 may be suggested for the biggest of them (Grangeiro, Luanda-Lameiro, Salamanca).

Finally, the head of the Carretão landslide, south of Crato, appears as a large-scale slump, below a 1.5 km long and 180 m high scar with smooth concave outline and profile (Fig. 5.11d). Below, a chaotic topography with 60–80 m high hills incised by four deep and narrow valleys disconnected from the scar extends over a width of 2 km down 540 m a.s.l. Formed by large tilted sandstone blocks in the upper part of the landslide, it is continued by a thinner debris apron that was emplaced over a differentiated topography, possibly at high speeds, as shown by ascending movements (Fort and Peulvast 1995), forming a debris avalanche or Sturzstrom (Angeli et al. 1996). The estimated area of the deposit is ~ 10 km^2 at minimum. The mean thickness probably represents several tens of meters in the slump area, and at least several meters (locally 10 m or more in the valleys) in the debris apron. These conservative figures indicate a minimal volume of 100 Mm^3.

Mass movements are also involved in the evolution of many other scarp segments, at all scales (Peulvast and Bétard 2015). Around the eastern tip of the Chapada (A type: see Sect. 3.1.1), they are similar to those of Crato–Barbalha area, with similar sizes, and well identified landslide scars and tilted blocks transported on the highest parts of the pediments (Jamacaru, Serra da Mãozinha, Porteiras…). In the glint segments, to the west, they are widespread, forming systems of simple or multiple terraces identified along the scarp in promontories, box canyons and sandstone outliers. These benches correspond to tilted sandstone blocks having slid along listric faults parallel to the plateau rim, over the underlying pelitic rocks of the Araripina Formation. They indicate a generalized scarp destabilization which certainly contributes to the strong dissection of the west tip of the Chapada and to its dismantling in mesas and buttes. Locally, wide pieces of the plateau have slid "en bloc", forming inward tilted terraces and causing the apparent repetition of the locally stepped ledges of the main scarp (Salitre). Some of the most spectacular sets of mass movement landforms are observed in the north rim of the Serra Vermelha, deeply scalloped by wide erosion cirques whose semicircular walls are fringed by 4–5 superposed benches similar to those identified SE of Barbalha, and in short box canyons of the west tip, west of Araripina, where wide tilted blocks clearly appear between the plateau and the valley floor. Here, contrarily to the Cariri depression, this process is generally not related to active vertical erosion, except on the Piaui side, since the underlying basement plateau is hardly dissected. There is only an ongoing process of lateral denudation, owing to a conjunction of factors (lithological superposition, hydrogeology, weak dip) highly favorable to scarp retreat. The gently sloping river profiles are sufficient for an evacuation of the fine waste (clay, sand) produced by mass wasting and later rock disintegration.

The particular activity observed in the northeast Araripe scarp (slumps and Sturzstrom belonging to the category of deep-seated landslides) is clearly related to the local structural conditions. They are as favorable as those listed by Moeyersons et al. (2008) in the Tigray province of Ethiopia: a tabular structure with very weak dips and a superposition of layers with contrasting mechanical and hydrogeological properties. Such conditions are not so well characterized in the Ibiapaba glint, where the sandstone cap generally exhibits stronger dips, less favorable to scarp retreat (Peulvast and Vanney 2001), whereas the underlying basement is more heterogeneous and often more resistant, in spite of local deep weathering. In the northeast Araripe basin, the layers of gypsum of the Ipubi Formation, as well as the marls and clays of the Romualdo Member exposed at the base of the scarp probably play a decisive role as weak levels, together with the 20–80 m thick Arajara clayey sandstones that underlie the massive Exu sandstone. All slumps described in our study are seated in these layers. Although the total thickness of gypsum layers remains moderate (a few tens of meters at most) the particular role of this rock as a ductile level prone to shearing at relatively low temperatures (Barberini et al. 2005) is commonly recognized and may be involved in major destabilization of slopes when sufficiently thick.

Permanent or seasonal saturation by water of parts of this series is probably also involved in the occurrence of such phenomena. Two aquifers outcrop in the scarp zone, corresponding to the Arajara sandstone and the base of the Exu Formation, and to the Crato limestone, respectively. Below the first of them, clays, marls and gypsum layers form an aquiclude ("Santana aquiclude") as well as a thick plastic level, the surface of which, when exposed, generally appears chaotic, impracticable during the rainy season, and commonly affected by shallow, small-scale slumps in valleys and road cuts. Combined with the mechanical weakness of the underlying gypsum, these characteristics probably explain the importance of deep-seated mass movements all along the scarp, although we have also observed generalized mass wasting in the Exu sandstone cliff even in areas where these soft basal sediments are missing. In this case, the presence of fine sediments in basal layers of the Exu sandstone and the strong clayey weathering observed along joints and fractures are probably influential as well.

5.3.3 Late Cenozoic Climate Change, Hillslope Processes and the Last Stages of Scarp Evolutions

Correlated to these deposits or older, the Barreiras sediments record a relatively short part of the Neogene erosion history (22–16 Myear: early to middle Miocene). A shown higher (Sect. 5.1.2), their deposition does not imply a significant Neogene acceleration of erosion and uplift, neither does an investigation of possible neotectonic movements outside coastal sites.

Climatic changes induced numerous short-term fluctuations since the Miocene, essential for the regradation of low surfaces by etch-planation and their inward extension (Peulvast and Claudino Sales 2004; Peulvast et al. 2008). The relative increase in clastic supply from the hinterland is probably linked to a marked shift toward aridity at that time, as suggested by Harris and Mix (2002) from a study of the ratio of oxide minerals in the terrigenous sedimentation of the Ceará Rise. The identification of coarse alluvial fans on dissected pediments, below high mountain slopes bearing marks of deep denudation and of strong demolition of rock ledges, and the presence of thick gravel terraces along middle or lower river courses probably reflect the occurrence of periods favorable to widespread stripping of deep soil horizons and even to erosion of bare rock slopes and surfaces, leading to generalized, yet moderate downwearing, and local backwearing.

As suggested by the data published by Harris and Mix (2002), such events may have taken place in dry conditions with discontinuous vegetation cover, allowing the occurrence of debris flows and torrential floods, and they were followed by dissection stages in periods of more humid climate and/or of lower sea level. The diversity of combinations between these sequences, already suspected in the sedimentary history of the Barreiras and Tibau deposits, reflects a complex environmental history in the Late Cenozoic. Although the repeated Pleistocene sea level falls probably explain the final dissection of the piedmonts, this complexity and the short duration of climatic and eustatic events were not favorable to continuous regressive erosion toward the scarps, explaining the frequent disconnection of the main river trunks from the scarps, even those bearing sparse marks of active erosion, especially in the widest coastal and inner piedmonts. As shown by the position of deposits of various ages on pediments and exhumed surfaces close to the scarps (Fig. 5.1), backwearing, if any, was also limited on the same period, even at the head or at the knick-points of the scarce gorges connected to the parts of the drainage system entrenched into the low surface (Baturité).

5.3.4 Conditions of Recent Evolution of Hillslopes and Implications for Hazard Assessment

In the Araripe scarp, large-scale gravitational dynamics and mass wasting are involved at least in the last stages of scarp retreat above the Cariri depression. Probably responsible for earlier retreat stages, as suggested by the nature of pediment covers and their distribution in distal parts of the dissected glacis strips, they are correlative of the recent development of erosional reentrants and of scarp exaggeration above increasingly deep valleys and depressions. The preserved volumes of deposits only represent a small proportion of the scarp retreat, still to be quantified. Matrix outwashing and weathering of coarse elements prepare the final waste exportation, by fluvial transport of sand and finer elements to the Salgado and

Jaguaribe river system. The role of intense, short rainfall events repeated during the rainy season, and causing groundwater saturation in the aquicludes is probably involved. Compared to the drier, surrounding plains of the Sertão, the northeast edge of the Chapada do Araripe receives higher amounts and intensities of rainfall, which may be considered as a potential factor of landslide triggering. As a working hypothesis, and waiting for "absolute" ages of some of the deposits described here, we also suggest a possible influence of Late Quaternary climatic changes, with wetter periods (prolonged periods of high precipitation) possibly favoring a greater occurrence of large-scale mass movements, such as revealed by paleoclimatic studies on NE Brazil (Behling et al. 2000; Wang et al. 2004). So far, we are still unable to ascertain whether we have still periodically active, or inactive processes.

Owing to contrasted climatic conditions and generally strong weathering in basement rocks as well as in cover sediments, mainly on humid slopes exposed to the trade winds (Bétard 2007), present-day slope instabilities are not uncommon, locally leading to well-characterized processes, landforms and deposits of gullying and mass wasting. Among them, those of the Chapada do Araripe deserve a special mention, since the adjustment of river- and hillslope profiles does not seem to be achieved and, on the contrary, to depend on the delay in the vertical erosion of a sandstone threshold, in the context of a general Late Cenozoic dissection of the low "Sertaneja" surface (Fig. 5.12). Therefore, in this case, but also in most other

Fig. 5.12 The Missão Velha waterfall, eastern Cariri depression. Exaggeration and retreat of the Araripe scarp (visible in the background) are related to late dissection of the low erosion surface. Here, this process is controlled by the outcrop level of resistant sandstone layers which form the outer skirt of the sedimentary Cariri depression (Mauriti Formation) and a local base level hardly deepened by the only fluvial outlet of the depression, i.e., the Rio Salgado. *Photograph* J.P. Peulvast

investigated scarps and mountains (e.g. the Baturité and Maranguape massifs), hazards related to slope processes should never be neglected, although only small-scale events were recorded in the last decades (Bastos 2012), and times of recurrence of larger events are probably much longer than the historic times (Peulvast et al. 2011). This is precisely what makes them more dangerous.

References

Angeli MG, Gasparetto P, Menotti RM, Pasuto A, Silvano S, Soldati M (1996) Rock avalanche. In: Dikau R, Brunsden D, Schrott L, Ibsen M (eds) Landslide recognition: identification, movement and causes. Wiley, Chichester, pp 190–201

Arai M (2006) Agrande elevação eustática do Mioceno e sua influência na origem do Grupo Barreiras. Revista do Instituto de Geociências, USP. Geol. USP Sér. Cient., São Paulo 6(2):1–6

Assumpção M (1992) The regional intraplate stress field in South America. J Geophys Res 97 (8):11889–11903

Barberini V, Burlini L, Rutter EH, Dapiaggi M (2005) High-strain deformation tests on natural gypsum aggregates in torsion. In: Bruhn D, Burlini L (eds) High-strain zones: structure and physical properties. Geological Society, London, Special Publications 245, pp 277–290

Barreto AMF, Bezerra FHR, Suguio K, Tatumi SH, Yee M, Paiva RP, Munita CS (2002) Late Pleistocene marine terrace deposits in northeastern Brazil: sea-level change and tectonic implications. Palaeogeogr Palaeoclimatol Palaeoecol 179(1):57–69. doi:10.1016/S0031-0182 (01)00408-4

Barros SDS, Fonseca VP, Saadi A (1997) Estudo morfo-tectônico preliminary nos platôs da Formação Serra do Martins/Bloco Portalegre-Marins (RN). XVII Simp Geol do Nordeste, Fortaleza, Res. Exp., SBG – Núcleo Nordeste 15, pp 365–369

Bastos FH (2012) Classificação dos movimentos de massa no Maciço de Baturité (CE) e contribuções para estratégias de planejamento ambiental. Ph.D. thesis, Universidade Federal do Ceará

Behling H, Arz HW, Pätzold J, Wefer G (2000) Late Quaternary vegetational and climate dynamics in northeastern Brazil: inferences from marine core GeoB 3104-1. Quatern Sci Rev 19:981–994

Bertani RT, Costa IG, Matos RMD (1990) Evolução tectono-sedimentar, estilo estrutural e habitat do petróleo na Bacia Potiguar. In: Raja Gabaglia GP, Milani EJ (eds) Origem e evolução de bacias sedimentares. Petrobras, pp 291–309

Bétard F (2007) Montagnes humides au cœur du Nordeste brésilien semi-aride: le cas du massif de Baturité (Ceará). Apports d'une approche morphopédologique pour la connaissance et la gestion des milieux. Ph.D. thesis, Univ Paris-Sorbonne

Bezerra FHR, Vita-Finzi C (2000) How active is a passive margin? Paleoseismicity in northeastern Brazil. Geology 28(7):591–594

Bezerra FHR, Amaro VE, Vita-Finzi C, Saadi A (2001) Pliocene-quaternary fault control of sedimentation and coastal plain morphology in NE Brazil. J South Am Earth Sci 14:61–75

Bezerra FHR, Fonseca VP, Lima-Filho FP, Saadi A (2005) Liquefaction-induced structures in quaternary alluvial gravels and gravelly sediments, NE Brazil. Eng Geol 76:191–208

Bezerra FHR, do Nascimento AF, Ferreira JM, Nogueira FC, Fuck RA, Brito Neves BB, Sousa MOL (2011) Review of active faults in the Borborema Province, intraplate South America integration of seismological and paleoseismological data. Tectonophysics 510:269–290

Bigarella JJ (1975) The barreiras group in Northeastern Brazil. An Acad Bras Ciên 47:365–393

Bigarella JJ, Ab'Saber A (1964) Palaeogeographische und palaeoklimatische Aspekte des Kanozoikums in Südbrasilien. Z Geomorph NF 8:286–312

Bigarella JJ, Andrade GO (1964) Considerações sobre a estratigrafia dos sedimentos cenozóicos em Pernambuco (grupo Barreiras). Univ Recife Inst CiênTerra Arquivos 2:2–14

Braga APG, Passos CAB, Souza EM, França JB, Medeiros MF, Andrade VA (1981) Geologia da região Nordeste do Estado do Ceará. DNPM, MME, Brasilia

Carneiro CDR, Hamza VM, Almeida FFM (1989) Ativação tectônica, fluxo geotérmico e sismicidade no Nordeste oriental brasileiro. Rev Bras Geociências 19(3):310–322

Claudino Sales V (2002) Evolution géomorphologique de la zone côtière de Ceará (Nord-est du Brésil): du long terme au court terme. Ph.D. thesis, Univ Paris-Sorbonne

CPRM-Serviço Geológico do Brasil (2003) Atlas digital de geologia e recursos minerais do Ceará. Mapas na escala 1:500,000. Serviço Geológico do Brasil, CD Rom

De Castro DL, Barbosa VCF, Silva JBC, Medeiros WE (1997) Relevo da interface crosta-manto entre os vínculos de isostasia e suavidade. Congr. Int. SBGf 5, São Paulo, Anais, pp 682–685

De Castro DL, Bezerra FHR, Sousa MOL, Fuck RA (2012) Influence of Neoproterozoic tectonic fabric on the origin of the Potiguar Basin, northeastern Brazil and its link with West Africa based on gravity and magnetic data. J Geodyn 54:29–42

Destro N, Szatmari P, Ladeira EA (1994) Post-Devonian transpressional reactivation of a Proterozoic ductile shear zone in Ceará NE Brazil. J Struct Geol 16(1):35–45

Ferreira JM, Assumpção M (1983) Sismicidade do Nordeste do Brasil. Rev Bras Geofis 1:67–88

Ferreira JM, Oliveira R, Takeya MK, Assumpção MA (1998) Superposition of local and regional stresses in NE Brazil: evidence from mechanisms around the Potiguar marginal basin. Geophys Journ Int 134:341–355

Fonseca VP, Melo FTL (1997) Análise morfo-tectônica da folha Milagres, extremos SE do Ceará e W da Paraíba. XVII Simp Geol Do Nordeste, Fortaleza, Anais, pp 383–386

Fort M, Peulvast JP (1995) Catastrophic mass movements and morphogenesis in the Peri-Tibetan ranges: examples from west Kunlun, East Pamir and Ladakh. In: Slaymaker O (ed) Steepland geomorphology. Wiley, Chichester, pp 171–198

Galvão CC, Da Silva FCA (2002) Análise microestrutural de um corpo siliciclástico localizado a Nordeste de Santana do Acaraú – CE. XLI Congr Bras Geol, João Pessoa, Anais, p 624

Gunnell Y (1998) Passive margin uplifts and their influence on climatic change and weathering patterns of tropical shield regions. Global Planet Change 18:47–57

Gurgel SPP, Bezerra FHR, Corréa ACB, Marques FO, Maia RP (2013) Cenozoic uplift and erosion of structural landforms in NE Brazil. Geomorphology 186:68–84

Harris SE, Mix AC (2002) Climate and tectonic influences on continental erosion of tropical South America, 0–13 Ma. Geology 30(5):447–450

Jardim de Sá EF, Matos RMD, Morais Neto JM, Saadi A, Pessoa Neto OC (1999) Epirogenia Cenozózoica na Província Borborema : síntese e discussão sobre os modelos de deformação associados. VII Simp. Nac. Estudos Tect., Lençoís – Bahia, Anais, pp 58–61

Jardim de Sá EF, Souza ZS, Vasconcelos PM, Saadi A, Galindo AC, Lima MG (2005) Marcos temporais para a evolução Cenozóica do Planalto da Borborema. Proceedings of the X Simposio Nacional de Estudos Tectônicos, Curitiba, Brasil, Boletim de Resumos Expandidos, pp 160–162

Keller EA, Pinter N (1996) Active tectonics. Earthquakes, uplift and landscape. Prentice Hall, Upper Saddle River, NJ

Klein C (1993) Du dynamisme des processus à la dynamique des formes en, géomorphologie edn. Ophrys, Gap

Leeder MR, Gawthorpe RL (1987) Sedimentary models for extensional tilt-block/half-graben basins. In: Coward MP, Dewey JF, Hancock PL (eds) Continental extensional tectonics, Geological Society Special Publication 28, pp 139–152

Lima MG (2008) A História do Intemperismo na Província Borborema Oriental, Nordeste do Brasil. Implicações paleoclimáticas e tectônicas. Ph.D. thesis, Universidade Federal do Rio Grande do Norte, Natal

Maia RP, Bezerra FHR (2014a) Tópicos de geomorfologia estrutural. Nordeste brasileiro, Edições UFC, Fortaleza

Maia RP, Bezerra FHR (2014b) Inversão neotectônica do relevo na Bacia Potiguar, Nordeste do Brasil. Revista Brasileira de Geomorfologia 15(1):61–74

Matos RMD (1992) The Northeast Brazilian rift system. Tectonics 11(4):766–791

Menezes MRF, Fonseca VP, Saadi A (1997) Morfo-tectônica na Serra de Santana (RN) e correlações com as estruturas regionais. XVII Simp Geol do Nordeste, Fortaleza, Res Exp, SBG – Núcleo Nordeste, 15:396–400

Moeyersons J, Van Den Eeckhaut M, Nyssen J, Tesfamichael G, Van de Wauw J, Hofmeister J, Poesen J, Deckers J, Haile M (2008) Mass movement mapping for geomorphological understanding and sustainable development: Tigray, Ethiopia. Catena 75:45–54

Morais Neto JM, Green PF, Karner GD, Alkmim FF (2008) Age of the Serra do Martins Formation, Borborema Plateau, northeastern Brazil: constraints from apatite and zircon fission track analysis. Bol Geoci Petrobras 16(1):23–52

Nascimento DA, Gava A, Pires JL, Teixeira W, Franfa FAB, Cabral EMA, Stamford WJP, Ribeiro AG, Barros MJG, Fonseca RA (1981) Geologia. In: ProjetoRadambrasil, levantamentos de recursos naturals, MME, vol 21. DNPM, Folha SA 24, Fortaleza, pp 23–212

Nóbrega MA., Bezerra FHR, Neto, JH, Iunes PJ, Guedes S, Tello Saenz CA, Hackspacher PC, Lima-Filho FP (2005) The use of apatite fission track thermochronology to constrain fault movements and sedimentary basin evolution in northeastern Brazil. Rad Meas 39:627–633

Nogueira FC, Bezerra FHR, Fuck RA (2010) Quaternary fault kinematics in intraplate northeastern Brazil. J Geodyn 49:79–91

Passos E, Bigarella JJ (1998) Superfícies de erosão. In: da Cunha SB, Guerra AJT (eds) Geomorfologia do Brasil. Bertrand Brasil, Rio de Janeiro, pp 107–141

Pessoa Neto OC (2003) Estratigrafia de seqüências da plataforma mista neogênica na Bacia Potiguar, margem equatorial brasileira. Rev Bras Geociências 33:263–278

Peulvast JP, Bétard F (2013) Late Cenozoic and present-day hillslope erosion dynamics in a passive margin context: stability or instability? Case studies in Northeast Brazil. Geografia Fisica e Dinamica Quaternaria 36(1):139–149. doi:10.4461/GFDQ.2013.36.0

Peulvast JP, Bétard F (2015) A history of basin inversion, scarp retreat and shallow denudation: the Araripe Basin as a keystone for understanding long-term landscape evolution in NE Brazil. Geomorphology 233:20–40. doi:10.1016/j.geomorph.2014.10.009

Peulvast JP, Claudino Sales V (2004) Stepped surfaces and palaeolandforms in the northern Brazilian "Nordeste": constraints on models of morphotectonic evolution. Geomorphology 62:89–122

Peulvast JP, Claudino Sales V (2005) Surfaces d'aplanissement et géodynamique. Géomorphologie Relief Processus Environ 4:249–274

Peulvast JP, Vanney JR (2001) Géomorphologie structurale, tome 1. Relief et structure. Gordon & Breach, BRGM Editions, Paris, Orléans

Peulvast JP, Claudino Sales V, Bezerra FHR, Bétard F (2006) Landforms and neotectonics in the equatorial passive margin of Brazil. Geodin Acta 19(1):51–71. doi:10.3166/ga.19.51-71

Peulvast JP, Claudino Sales V, Bétard F, Gunnell Y (2008) Low post-Cenomanian denudation depths across the Brazilian Northeast: implications for long-term landscape evolution at a transform continental margin. Global Planet Change 62:39–60. doi:10.1016/j.gloplacha.2007.11.005

Peulvast JP, Bétard F, Magalhães AO (2011) Scarp morphology and identification of large-scale mass movements in tropical tablelands: the eastern Araripe basin (Ceará, Brazil). Géomorphologie: relief, processus, environnement 1:33–52. doi:10.4000/geomorphologie.8800

Rossetti DF (2004) Paleosurfaces from Northeastern Amazonia as a key for reconstructing paleolandscapes and understanding weathering products. Sed Geol 169:51–174

Rossetti DF, Bezerra FHR, Góes AM, Brito Neves BB (2011) Sediment deformation in Miocene and Post-Miocene strata, Northeastern Brazil: Evidence for paleoseismicity in a passive margin. Sed Geol 235:172–187

Rossetti DF, Bezerra FHR, Dominguez JLR (2013) Late Oligocene-Miocene transgressions along the equatorial and eastern margins of Brazil. Earth-Sci Rev 123:87–112

Saadi A, Torquato JR (1992) Contribuição à neotectônica do Estado do Ceará. Rev Geologia 5:5–38

Silva CG (1991) Holocene stratigraphy and evolution of the Açu River Delta, Rio Grande do Norte State, Northeastern Brazil. Ph.D. thesis, Duke University

Sousa DC, Jardim de Sá EF, Antunes AF (2008a) Deformação neógena e suas implicações na estruturação dos campos de petróleo na região de Icapuí-Ponta Grossa (CE), Bacia Potiguar emersa. Revista Brasileira de Geologia 38(2-suplemento):97–110

Sousa DC, Jardim de Sá EF, Vital H, do Nascimento MAL (2008b) Falésias na praia de Ponta Grossa, Icapuí, CE. Importantes deformações tectônicas cenozóicas em rochas sedimentares da Formação Barreiras. In: Winge M, Schobbenhaus C, Souza CRG, Fernandes ACS, Berbert-Born M, Queiroz ET (eds) Sítios geológicos e paleontológicos do Brasil. Published in Internet 24 Sept 2008. http://www.und.br/ig/sigep/sitio120.pdf

Suguio K, Bidegain JC, Mörner NA (1986) Dados preliminares sobre as idades paleomagnéticas do grupo Barreiras e da Formação São Paulo. Revista Brasileira de Geociências 16:171–175

Takeya MK, Ferreira JM, Pearce RG, Assumpção M, Costa JM, Sophia CM (1989) The 1986–87 intraplate earthquake sequence near João Câmara, notheast Brazil—evolution of seismicity. Tectonophysics 167:117–131

Teixeira AL, Gaucher C, Paim PSG, Fonseca MD, Parente CV, Silva Filho WD, Almeida AD (2004) Bacias do estágio da transição da Plataforma Sul-Americana. In: Bartorelli A, Carneiro CDR, Brito-Neves BB (eds) Mantesso Neto V. Geologia do Continente Sul-Americano, Evolução da Obra de Fernando Flávio Marques de Almeida, Beca, pp 487–537

Torquato JR, Torquato AM, Moraes RM (1991) O graben da Fazenda Retiro Grande e outros aspetos neotectônicos na região da Praia de Redonda (Ceará). Geonomos 4(2):17–21

Wang XF, Auler AS, Edwards RL, Cheng H, Cristalli PS, Smart PL, Richards DA, Shen CC (2004) Wet periods in northeastern Brazil over the past 210 kyr linked to distant climate anomalies. Nature 432:40–43

Chapter 6
Conclusion and Perspectives

Abstract The discontinuous escarpment of Ceará-Rio Grande do Norte appears as a composite landform. Its definition matches that of Great Escarpments that develop on initial high-elevation margins, with landward retreat of the initial rift-flank uplift. Nevertheless, the term "marginal escarpment" appears more appropriate, corresponding to a particular tectonic and geodynamic pattern (transform margin without strong uplift). In all parts of the marginal escarpment, significant erosion, probably more by downwearing than by backwearing, took place, helped by post-rift flexure that reduced the thickness of the rock slice to be eroded. Our estimation of regional magnitudes of Cenozoic crustal deformation and denudation leads us to discuss recently published models of long-term geomorphic evolution based upon the results of thermochronological analyses. We stress the fundamental importance of taking into account the results of geomorphic approaches in any interpretation of thermochronological and other analytic methods used for reconstructing long-term landscape evolutions. Moreover, the morphostructural and morphopedological approaches used in our study of landforms and landscape evolution in Northeast Brazil gives some clues and new scientific data that help assessing geodiversity of this large territory. Our contribution relies both on geomorphodiversity and pedodiversity aspects. At a finer scale, it may also help identifying valuable geomorphological sites (or geomorphosites) whose scientific and other values (cultural, aesthetic, ecological…) might justify their integration into the national network of Brazilian geosites.

Keywords Geodiversity · Geomorphodiversity · Georisks · Geotourism · Long-term evolution models · Marginal escarpment · Pedodiversity · Thermochronology

J.-P. Peulvast and F. Bétard, *Landforms and Landscape Evolution of the Equatorial Margin of Northeast Brazil*, Springer Earth System Sciences,
DOI 10.1007/978-3-319-18203-2_6

6.1 Structural Geomorphology, Morphostratigraphy and Morphopedology in Northeast Brazil: New Ideas on Landforms and Landscape Evolution

6.1.1 Great Escarpment or Marginal Escarpment?

The study area is part of the Brazilian Equatorial continental margin, formed in Aptian times. Onshore, the Precambrian basement is organized around NE–SW shear zones, some of which were reactivated by preopening rifting (Cariri-Potiguar rift zone) and then intersected by the transform passive margin (Potiguar basin area; see Chap. 2). A thin dissected layer of Cenozoic sediments (Barreiras Group) is preserved over a 10–80 km wide coastal strip, forming low-lying plateaus or "*tabuleiros*" (10–60 m) at the foot of two offset segments of a discontinuous marginal scarp. The slightly dissected coastal lowlands merge inwards with vast corridors and plains (the Sertaneja surface) that extend far into the semiarid interior between flat or rugged plateaus and ridges of moderate altitudes (800–1200 m), shaped into a wide flexural uplift zone.

The crest zone of this belt of highlands exhibits relatively constant elevations between 1000 and 1200 m a.s.l. and forms the drainage divide between the São Francisco watershed and other rivers among which the Jaguaribe and Piranhas are the most important. Shorter drainage systems (Rio Acaraú, Baturité area, Rio Apodi) reflect the establishment of secondary divides within the deep topographic embayment defined by the escarpment. These divides correspond to a collection of isolated massifs (e.g., Meruoca, Baturité, Pereiro, Central Ceará highlands) that share similar altitudes (800–1000 m). Their steep northern termination outlines two discontinuous and roughly E–W alignments of elevated topography (Ibiapaba–Meruoca–Irauçuba–Baturité, and Pereiro–Santana, respectively: see Figs. 3.1 and 3.14), parallel to the continental margin. They correspond to the north extremities of the rift shoulders that surround the Cariri-Potiguar rift zone—respectively the north Ibiapaba–Uruburetama–Baturité alignment and the north Borborema-Serra de Santana promontory, but a part of the easternmost segment is also formed by inner structures of this rift zone, i.e., eroded and discontinuous remnants of the southern Potiguar rift shoulder. The offset zone between the two segments, which includes the widest plains and low plateaus of the study area—i.e., the Choró-Jaguaribe plain and the Chapada do Apodi—is also widely related to rift structures and areas of post-rift subsidence (the Potiguar basin and the eastern part of the Aracati platform). However, no direct tectonic control of the escarpment outlines is observed. Erosional retreat or dismantling of former rift shoulders from remote structures must be considered in most cases.

Therefore, the discontinuous escarpment of Ceará-Rio Grande do Norte appears as a composite landform. The Parnaiba–Fortaleza and Aracati–Macau alignments of residual highlands (Fig. 3.1) form a topographic limit between the inner highland region and a coastal piedmont partly covered by Cenozoic and older sediments. As such, they correspond to the eroded remains of Cretaceous rift flanks now deeply breached by embayed drainage basins and forming residual buttes, suggesting that the NE Brazilian plateau edge once extended as far north as these alignments. The Pereiro–Santana alignment corresponds to the eroded stumps of the Neocomian Potiguar half-graben rift shoulder. The other alignment south and west of Fortaleza would represent the eroded remains of a somewhat younger rift shoulder linked to transtensional opening of basins before the formation of the Equatorial Atlantic Ocean (see Fig. 2.5). The later evolution is related to almost continuous subsidence and deposition in the offshore Potiguar basin and to the complex evolution of the Ceará basin, characterized by moderate and discontinuous Tertiary sedimentation. In both cases, these escarpments appear to be widely inherited from Cretaceous tectonics, even if later flexural uplift and erosional modifications of their shapes and outlines probably occurred.

The definition of the resulting escarpment matches that of Type-1 Great Escarpments of Pazzaglia and Gardner (2000), i.e., those that develop "on initial high elevation margins where the rift-flank escarpment is maintained by minimal post-rift offshore sediment loading and flexural subsidence," with landward retreat of the initial rift-flank uplift (Kooi and Beaumont 1994). Here, a significant part of landward retreat would have occurred in the Cretaceous, as shown south of the Potiguar basin. Nevertheless, the term "Great Escarpement" does not reflect the discontinuous outlines of this escarpment and its moderate altitudes. Therefore, the term "marginal escarpment" appears more appropriate for describing such short escarpments or alignments of small mountain ranges or ridges that are found along a few other passive margins, intersecting or not oblique or transverse fault and rift patterns (e.g., British Islands, Maritime Provinces of Canada, Guyana, Guinea, Southwest Australia…). More than merely being transition forms between Great Escarpments and passive margins of low altitude (Argentine passive margin: Brown et al. 2000; Bétard et al. 2014; French Atlantic margin: Peulvast and Vanney 2002; Bétard 2010; south coast of Southwest Australia: Beard 2003), they correspond to particular tectonic and geodynamic patterns where a particular context (here, that of a transform margin without strong uplift) must be considered.

6.1.2 Stepped Surfaces and Morphostratigraphic Relationships: A Summary

The morphological patterns of northeast Brazil are classically considered as a result of regional uplift induced by breakup of the Gondwana supercontinent, interfering with climatic fluctuations in the Cenozoic (Mabesoone and Castro 1975). The uplift would have triggered the shaping of successively lower planation surfaces and the incision of a centrifugal drainage pattern (see Fig. 1.5).

Our chronology of the stepped surfaces identified in the regional landscape is far from matching those of classical models. In the south, the Araripe–Borborema region bears the highest well-dated surfaces, the Cenomanian fill- (and structural) surface and its beveled surroundings, incorporating at slightly different levels some elements of exhumed pre-Albian and Lower Paleozoic planation surfaces. North of the Patos fault zone, two different patterns are identified on both sides of the Senador Pompeu fault zone (Peulvast et al. 2008; see Fig. 4.7). On its southeast side, in the Cariri–Potiguar rift zone, the Sertaneja surface is widely developed at low altitudes around Cretaceous grabens, extending far inland along the main rivers, with low gradients, up to the Araripe basin. Deeply inset below remnants of high surfaces, it is clearly post-Cretaceous in age, but it coincides with the exhumed pre-Cretaceous surface in its distal parts. The polygenetic Sertaneja surface is buried near the coast, together with the post-Campanian surface of the Chapada do Apodi, below the Barreiras sediments.

To the northwest, no identified element of a Cretaceous surface is found at low altitude, but Cretaceous faulting is locally suspected to have facilitated its development (Acaraú plain). The Sertaneja surface is best developed in the vicinity of the Senador Pompeu fault zone, around the Baturité massif, and along the oceanic façade, forming large embayments between high dissected mountain blocks (Canindé, Santa Quiteria). Its development is clearly related to the base levels formed by the Jaguaribe plain to the east and the ocean to the north. Probably reworked until its Late Tertiary or Pleistocene dissection, but widely developed before the deposition of the Barreiras sediments in its distal parts, it may be considered as resulting from cyclic development of a "young" surface below remnants of poorly dated high surfaces.

6.1.3 Differential Erosion, Long-Term Landform Evolution and Morphopedology

6.1.3.1 The Decisive Role of Differential Erosion Responding to Regional Uplift

Our study stresses the decisive role of differential erosion in the landscapes of the study area, in spite of many uncertainties on the persistence of residual tectonic landforms or on a possible role of neotectonics. However, the available lithological data only give a part of the numerous parameters involved in the resistance of rocks to selective erosion. Many landforms of differential erosion are described in interior regions, in the basement and at the contact between basement and sedimentary covers, in places where direct tectonic control on landforms can be excluded (dissected erosion or structural surfaces, for instance). Their existence is confirmed by recent ^{10}Be analyses on rock outcrops within the Sertaneja depression (Seridó valley, RN), which show moderate (4–11 m/Myear) erosion rates, with lower rates (<3 m/Myear) on quartzite domes and slightly higher rates on micaschists (3–7 m/Myear) and pegmatites (8–13 m/Myear) (Morais Neto et al. 2010).

Once again, the morphology of the piedmonts and surroundings of the Chapada do Araripe illustrates the contribution of our study to this fundamental thematic of geomorphology. It reflects different conditions of preservation of exhumed surfaces or development of new erosion topographies during the regional inversion process which was shown as the main factor of geomorphic history of the region. Since differential erosion responding to regional uplift is the main mechanism of relief inversion (Peulvast and Bétard 2015), lithological contrasts between the basement and the overlying series are the main factors of this differentiated evolution, as well as the outcrop conditions controlled by the various unconformities observed throughout the basin. From our geological profiles, it appears that very few exhumed surfaces remain intact and that original post-Cenomanian landform shaping was very efficient, in particular in easily weathered metamorphic rocks such as micaschists, whereas the sandstone cap tended to protect the underlying rock surfaces from erosion until it was dismantled.

As shown in our geomorphic review of a large set of scarps and hillslopes within the northern regions of northeast Brazil (Peulvast and Bétard 2013), backwearing processes (sapping, landslides) involved in the dismantling of this sandstone cap are much less common in basement areas. The morphology of the lower steps formed around the Chapada do Araripe after exhumation of the basement confirms this observation, in particular for the lower step, which may correlate with the Sertaneja Surface or its equivalents on the north as on the south side. North of the Cariri depression, the exact coincidence between this surface and a micaschist strip shaped as a pediment now dissected in convex hills suggests that planation processes were limited to these weakly resistant rocks, whereas orthogneiss remained salient to the north. This is confirmed by the exhumation of granitic residuals. More generally, the complexity of local structural controls that fixed scarp and residual outlines in the basement is best explained in a context of evolution by etch-planation and downwearing (Thomas 1989a, b) limited to the most easily weathered rock volumes. The wide pediment entrenched below the sandstone plateau south of Exu is most regular in micaschists, whereas it only cuts the periphery of the granitic intrusion which underlays the sandstone cap. Here, approximate parallelism is observed between the scarp and the exhumed contact between granite and micaschist: it is therefore probable that limited scarp retreat (a few km at most) occurred into the granite intrusion during the cyclic entrenchment of this pediment by downwearing.

6.1.3.2 Regional Predominance of Paleolandforms and Differential Erosion Features: Downwearing Versus Backwearing

The northeast Brazilian landscape was shown to be the result of an episodically fluctuating but falling base level since the Middle Cretaceous, and crustal deformation in the form of a broad monocline between the interior Araripe and coastal Potiguar basins without any logical requirement or field evidence of significant fault offsets outside a few coastal areas. After formation of the passive margin, landscape development was driven by a swell-like uplift with its crest situated ~300 km from the coastline. Resulting long-term denudation rates in this setting were <10 m/Myear. These low post-rift denudation rates are explained by a conjunction of four factors which also influence the conditions of scarp formation and evolution: (i) the low magnitude of crustal uplift estimated by the current elevation of marine Albian layers (Araripe, Apodi); (ii) the low amplitude and long wavelength of crustal deformation of an initially low-relief topographic surface, which promotes a phenomenon defined as 'morphological resistance' (Brunsden 1993a, b) in which the development of high angle slope systems favorable to intense erosion is impeded; (iii) the lithological heterogeneity of the basement and its cover, with resistant bedrock outcrops (e.g., sandstone, limestone, granite) explaining the widespread preservation of residual topography still upstanding on the Sertaneja erosional plain and occupying over ~50 % of its surface area; (iv) the long-term semiarid climate in NE Brazil, probably in existence for the last 13 Myear at least (Harris and Mix 2002).

The seaward flank of this swell was eroded and currently forms the deeply embayed Sertaneja plain girdled by a semicircular erosional escarpment. The latter is partly controlled by lithological discontinuities between the basement and overlying siliciclastic cover rocks, which were enhanced by post-Cenomanian basin inversion. Post-Cenomanian denudation involved rates <10 m $Myear^{-1}$ over vast areas and limited post-rift retreat of existing scarps. In the process, denudation has exhumed a number of Cretaceous stratigraphic unconformities, so that some topographic surfaces at low elevations are effectively ancient land surfaces that became re-exposed in Cenozoic times. This peculiarity is reminiscent of the Australian continent and parts of Africa, where limited magnitudes of basement uplift (epeirogeny) during the Cenozoic have promoted shallow stripping of marine Cretaceous cover rocks. The predominantly clastic Barreiras and Tibau Formation testifies to some of the last stages of erosion in the hinterland followed by the onset of more arid climates at ~13 Myear or earlier.

Inland, although no clear evidence of recent tectonic movement is identified along outstanding geomorphic features, most faults along or near which recent seismotectonic activity is recorded control at least some important features of the regional relief. Many of them are oblique or perpendicular to the coastline and correspond to Precambrian shear zones reactivated during the Cretaceous rifting: the Transbrasiliano, Senador Pompeu, Jaguaribe, and Portalegre-Carnaubais shear zones. Some of these (Transbrasiliano and Senador Pompeu) are suspected to play major roles in massive decoupling between compartments of the margin that have undergone different post-breakup tectonic evolutions. Even in places where a possible influence of neotectonic movements is suspected (e.g., along the Serra do Pereiro), the respective roles and rhythms of erosive shaping (downwearing, backwearing in the uplifted block) and of uplift generally remain poorly known, making ambiguous their exact significance. On the contrary, the Potiguar basin is clearly bounded to the south by a residual fault scarp that was formed and has retreated from the Carnaubais and Apodi faults.

In all parts of the marginal escarpment, significant erosion, probably more by downwearing than by backwearing, took place, helped by post-rift flexure that reduced the thickness of the rock slice to be eroded. This situation is that of all the borders of inner and coastal surfaces forming the Sertaneja plain. It explains the high sinuosity of many scarps and the development of wide embayments,

Fig. 6.1 The Ibiapaba glint at Ubajara (NW Ceará). Here, the glint reaches its maximum height (600–700 m). Short dissected concave slopes and pediments, only interrupted by scattered limestone hills (Freicheirinha Formation), are shaped below the abrupt sandstone cornice (Serra Grande sandstones). The sandstones surmount thick and deeply weathered deposits of the Ubajara Group (clastic rocks and carbonates, metamorphosed to upper greenschist facies). Differential erosion related to deep dissection of the exhumed infra-Paleozoic surface preserved thick limestone or marble lenses, forming prominent karstified hills or towers (Ubajara) below the glint. *Photograph* J.P. Peulvast

frequently controlled at small scale by lithological contacts and tectonic weakness zones. It implies old ages (Cretaceous to Tertiary) for most of these landforms. In several places, especially below the Paleozoic paleosurface, deep weathering of other metamorphic rocks largely prepared a selective removal of the surrounding material, resulting for example in the emergence of marble buttresses (Ubajara, Fig. 6.1): a good illustration of the fact that backwearing of the escarpment from which they progressively emerge is often the manifestation of downwearing in localized form (Pugh 1966).

6.1.3.3 The Contribution of Morphopedology

The regional distribution of soils is highly reflective of the stepped pattern of landforms and associated denudation rates (Bétard and Peulvast 2013). Whereas only non-lateritic soils developed from shallow, grus-type weathering mantles occur on the Sertaneja surface marked by semiarid conditions since the Miocene, relict laterites inherited from older (Paleogene) paleoclimates remain currently scarce in the landscape and are confined to ancient summit surfaces and/or in positions of major drainage divides (Serra do Pereiro, Araripe–Campos Sales area). This organization of soil covers is symptomatic of contrasted denudation rates, with higher rates on the low-altitude Sertanjea surface steering the landscape toward the stripping of most of the paleoweathering mantle, and favoring the development of a primary (non-lateritic) fersiallitic pedogenesis under the newly established semiarid conditions, in existence at least since the last 13 Myear (Harris and Mix 2002; Bétard et al. 2009; Bétard 2012).

6.1.4 A Discussion of Long-Term Evolution Models

Our estimation of regional magnitudes of Cenozoic crustal deformation and denudation lead us to discuss recently published models of long-term geomorphic evolution based upon the results of thermochronological analyses. The lack of any remnant of post-Exu sediments in the southern part of the study area, including in the post-rift series of the Jatobá-Tucano basin (Magnavita et al. 1994), suggests that a younger sedimentary cover never existed there, in spite of recent interpretations of thermochronological results according to which a slice of 1000 m or more would have existed over the present pile (Morais Neto et al. 2005–2006; Japsen et al. 2012), before being totally removed.

Indeed, a post-Cenomanian deposition would be in weak concordance with the interpretation given by Harman et al. (1998) and confirmed by Morais Neto et al. (2005–2006), of a thermal history indicating a cooling episode beginning between 100 and 90 Myear (late Albian–Turonian), i.e., toward the end of the fluvial sedimentation, and a later one in the Oligocene. The first cooling episode would correspond to an exhumation stage which seems hardly compatible with then ongoing sedimentation.

The other difficulty encountered with the "episodic burial and exhumation model" developed by Japsen et al. (2012) for the post-rift evolution of the Brazilian Northeast, relates to the origin of the additional overburden (1000 m or so) inferred from AFT analysis. The corresponding volume would have been huge, and its source remains highly uncertain. We have shown this is highly improbable (Sect. 4.3.3). As demonstrated higher, the exhumation and partial destruction of paleosurfaces probably began early around the Chapada do Araripe, during the Paleogene. Here, these early stages of topographic inversion might correspond to the first post-rift denudation episode reported by AFT. The second stage, in the Oligocene and later, would correspond to the major phase of river incision and basin inversion. Responding to regional uplift, the differential erosion is also controlled by retreat of the river knickpoints proceeding from the developing low-lying Sertaneja Surface. It reached the lowermost post-rift layers, cutting into the syn-rift formations and even into the underlying basement outcrops of the former rift shoulders. The corresponding mean denudation rate (7–10 m $Myear^{-1}$) is similar to that of inferred vertical movements.

Therefore, we stress the fundamental importance of taking into account the results of geomorphic approaches in any interpretation of thermochronological and other analytic methods used for reconstructing long-term landscape evolutions. Elaborating “new” models that forget the reality of landforms and geological structures, the stratigraphic evidence or the distribution of soil covers only leads to improbable assumptions that weaken all attempts to understanding the regional landscape patterns. As a return, elaborating more realistic models of long-term landform evolution including denudation rates and modalities based upon morphostratigraphic analysis, when possible, will bring elements for a better calibration of the physical methods. This is the price to be paid for improving the trust one can have in their results, which bring complementary or unique information, peculiarly in places where sedimentary, volcanic or pedological markers are missing. Such a knowledge is also necessary for checking evaluations of the tectonic component in relief formation extrapolated from measures on small-scale features and short periods (Serra do Pereiro: Gurgel et al. 2013; Chapada do Araripe: Marques et al. 2014). All the new ideas exposed above concerning landforms and landscape evolution have strong implications for further fundamental researches in Northeast Brazil, but new challenges are also emerging in applied research in the field of geodiversity, in a region of rich geoheritage to be inventoried, protected and promoted.

6.2 Toward an Integrated Assessment of Geodiversity in Northeast Brazil

More recent and less known than the biodiversity concept, geodiversity can be defined as “the natural range (diversity) of geological (rocks, minerals, fossils), geomorphological (landforms, processes) and soil features. It includes their assemblages, relationships, properties, interpretations and systems” (Gray 2004). Whereas geodiversity refers to the variety of abiotic nature, geoheritage is the set of the most relevant geodiversity elements with particular importance for science, education or tourism (Pereira et al. 2012). During recent years, some attempts were made in order to assess geodiversity and to inventory geoheritage and geosites in the territories at various scales. Recently in Brazil, the National Geological Service (CPRM) published the Geodiversity Map of Brazil at 1:2,500,000 scale (CPRM 2006), which is a synthesis of the major geosystems constituting the national territory, as well as their limitations and potential uses for geotourism or georesource exploitation (Silva 2008). However, until now, the CPRM geodiversity map is only based on geological data such as lithostratigraphy and mineral resources and does not take into account other geodiversity elements such as landforms and soils, which are also important to interpret the whole geodiversity and to support decision-making and land-use management issues (Pereira et al. 2012).

6.2.1 *Geodiversity in Ceará and Neighboring Regions: A Preliminary Assessment*

6.2.1.1 The Contribution of Morphostructural and Morphopedological Approaches to the Knowledge of Geodiversity

The morphostructural and morphopedological approaches used in our study of landforms and landscape evolution in Northeast Brazil gives some clues and new scientific data that help integrating other elements of geodiversity than the geological diversity *stricto sensu* (as exposed in Chap. 2). Indeed, our contribution relies both on geomorphodiversity (morphostructural patterns, climatic landforms and palaeolandforms, morphodynamic features) and pedodiversity aspects (soils and palaeosoils, laterites and duricrusts).

The peculiarities of that region may be first described from a geomorphodiversity viewpoint (Panizza 2009), i.e., "the critical and specific assessment of the geomorphological features of a territory, by comparing them in an extrinsic and intrinsic way, taking into account the scale of investigation, the purpose of the research and the level of scientific quality." At larger scale (i.e., continental scale), the morphostructural patterns described in Chap. 3 reveal the great variety of structural landforms in various geological contexts (sedimentary, crystalline, volcanic…), that points out a high degree of extrinsic geomorphodiversity compared to nearby platform regions (e.g. the monotonous flat landforms and landscapes of the inner Parnaiba basin, in the Piaui state). The coexistence of typical granitic landforms (inselbergs, bornhardts, tors and microforms of the Quixadá granitoid complex), karstic landforms on limestone and sandstone caprocks (karren, giant lapiaz and caves of the "Chapada" landscapes) and volcanic landforms (necks, dykes and plugs of the Oligocene Messejana volcanism) make this region an open-air museum of lithology-dependent landforms which have a great value from an educational viewpoint (e.g., Migoń 2006). The stepped pattern of palaeolandforms of various ages (Paleozoic, Mesozoic or Cenozoic: see Chap. 4) greatly accentuates extrinsic geomorphodiversity: indeed, the present relief reflects a juxtaposition of highly contrasted elements of varied ages, identified by the means of a morphostratigraphic analysis of several types of unconformities, erosion surfaces, palaeoescarpments, and palaeopiedmonts (Peulvast and Claudino Sales 2004).

At regional scale, the present-day climatic conditions are highly contrasted in the Brazilian Northeast: although this region largely belongs to the semiarid Nordeste (or "Sertão"), the coastal strip and the inner mountains are characterized by more humid tropical climates, with forest environments, that necessarily influence the morphodynamics. This bioclimatic pattern typically contrasts two morphological types and climatic landforms in the regional setting (Bétard et al. 2007): a multiconvex topography with *meias laranjas* and flat valley bottoms in the humid zones (e.g., the summit surfaces of the Baturité and Meruoca massifs) and a multiconcave topography formed by mantled pediments and inselbergs in the semiarid areas of the "Sertão." This intrinsic geomorphodiversity is reinforced by the numerous morphoclimatic legacies induced by a complex paleoenvironmental history: the warm and humid environments of the Palaeogene and the alternation of wet and dry periods during the Neogene and the Pleistocene have generated a sequence of mosaic landforms shaped according to various processes (e.g., the dominant aeolian processes of the driest periods of the Pleistocene, responsible for the existence of complex palaeodune systems and the deposition of the Post-Barreiras sediments accumulated in a wide coastal strip: Bétard and Peulvast 2011). At finer scale (i.e., local scale), many geomorphological objects inventoried in this book (see Chap. 5) contribute to increase the intrinsic geomorphodiversity of the area as, for example, the great variety of landslides and other mass movements recently discovered in the rims of the Chapada do Araripe (Peulvast et al. 2011). Because of their high scientific and/or additional values (cultural, educational, touristic…), many of the sites described in this book are potential geomorphosites that might be further integrated into the national network of Brazilian geosites (SIGEP inventory: http://sigep.cprm.gov.br/).

In a complementary perspective, the morphopedological approach is an interesting way for addressing the emerging concept of pedodiversity that can be defined as the distribution, extent and variety of soils (Ibáñez et al. 1995; Tennesen 2014). Actually, few studies have considered pedodiversity (i.e., soil diversity) in the assessment of the wider, integrating concept of geodiversity. In addition to existing soil maps and inventories (e.g., Guichard 1970; RadamBrasil 1981a,b) that can help in evaluating the regional pedodiversity, our morphopedological observations in several parts of the Ceará state may also contribute to the qualitative assessment of pedodiversity considered here as a subfield of geodiversity. For instance, the detailed morphopedological study carried out in the Baturité massif and its piedmont (Bétard 2007, and the 1:250,000 morphopedological map therein) provides a basis for such a preliminary assessment at a regional scale. Three main soil–landform units related to the present-day bioclimatic pattern were effectively recognized and delineated (Bétard 2012; Fig. 6.2): (i) the humid massif is characterized by a summit surface (800–900 m a.s.l.) shaped into multiconvex topography supporting Acrisols on the convex hills, and Fluvisols with waterlogging features on the valley bottoms; (ii) the subhumid peripheral area is typified by highly dissected landscapes (200–800 m a.s.l.), with Lixisols in the hilltops and Leptosols on the eroded slopes of V-shaped valleys; (iii) the semiarid piedmont (100–200 m a.s.l.) exhibits a typical landscape of mantled pediments and inselbergs covered by Luvisols and Cambisols on wash divides, grading to Planosols and Fluvisols in valley bottoms. To the east, part of the semiarid piedmont extends the pre-littoral tablelands ("Tabuleiros") which are composed of basal Neogene clastic sediments covered by palaeosoils (Lixisols) and of upper Pleistocene aeolian sands underlain by weakly developed soils (Arenosols). In other parts of the study area, the inventory and systematic mapping of laterites and duricrusts are also key elements of pedodiversity and, implicitly, of geodiversity. Many ferruginous duricrusts are powerful indicators of long-term landscape evolution and, as such, have a great scientific value (e.g., the laterized band of the Campos Sales area, just below the Chapada do Araripe: Bétard et al. 2005; Peulvast and Bétard 2015). Their cultural and/or

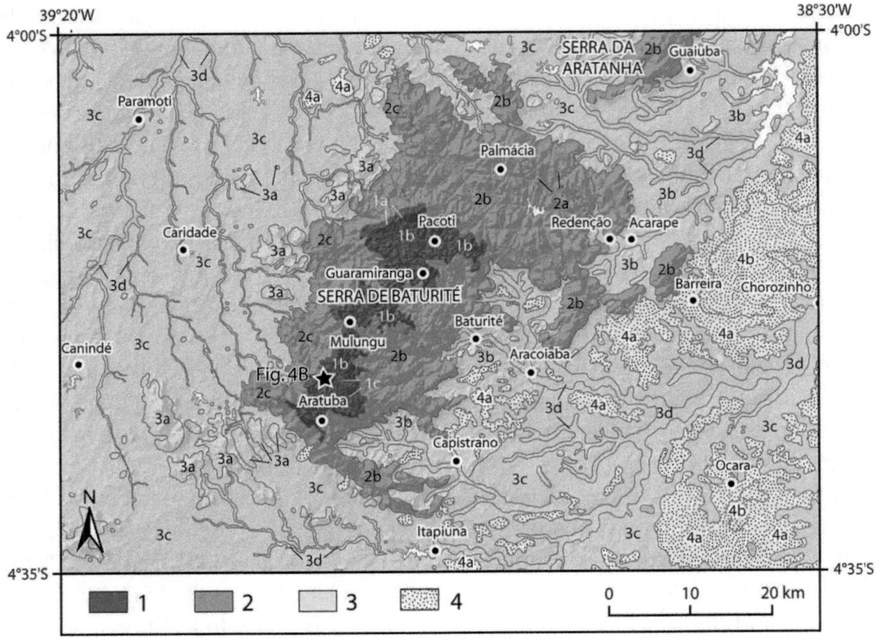

Fig. 6.2 Simplified morphopedological map of the Baturité massif and its piedmont (Northeast Brazil). This sketch corresponds to a simplification of the 1:250,000 scale morphopedological map edited by Bétard (2007). *1* Summit area of the Baturité massif with 'meia laranja' topography (Acrisols dominant); *2* peripheral area of highly dissected landscape with rocky domes (Lixisols dominant); *3* 'Sertão' plain with pediments and inselbergs (Luvisols dominant); *4* pre-littoral tablelands or 'Tabuleiros' (Arenosols dominant)

economic uses as disused quarry sites are additional values that make them potential geosites to protect and valorize, at least locally.

6.2.1.2 The Values of Geodiversity

Geodiversity constitutes a basis for the selection of geosites, which should be chosen to represent the geodiversity of a country, province or region (Gray 2004). To help in achieving this selection more objectively, over 30 values of geodiversity have been identified by M. Gray based on: (i) intrinsic value (free of human valuation), (ii) cultural values (e.g., legends or religious concerns); (iii) aesthetic values (e.g., privileged scenic views and impressive landscapes), (iv) economic values (e.g., coal, geomaterials, metals, gemstones); (v) functional values (e.g., geodiversity as a support for biodiversity); (vi) scientific/educational values (e.g., geoeducation for scholars). In other terms, these values can be viewed as 'geosystem services', the abiotic equivalent of the 'ecosystem services' that have been debated so far by ecologists and economists to demonstrate the value of biodiversity for human society and well-being (Gray 2008).

Considering the implicit scientific interest of many geosites and geomorphosites of Northeast Brazil described in this book, it is assumed they all have an intrinsic value which may justify their integration to a regional or national network of geosites, with respect to their level of scientific interest (from local to international). In the Araripe Geopark located in southern Ceará—the only Brazilian representative in the Global Network of Geoparks—, Mochiutti et al. (2012) have identified additional values (cultural, aesthetic, economic, functional, educational) in each geosite. Many potential geomorphosites of Northeast Brazil also have such additional values. The cultural value is pregnant in the Quixadá geomorphosite (Fig. 6.3a) and the Sete Cidades National Park (Fig. 6.3b), mainly associated with Kariri prehistoric settlements and rupestrial paintings. The aesthetic value is particularly significant in panoramic geomorphosites such as the Pontal da Santa Cruz in the Araripe Geopark (Fig. 6.3c), and in coastal geomorphosites such as the Jericoacoara geodiversity hotspot (Fig. 6.3d). The economic value appears in sites of traditional mining and exploitation of geomaterials as, for example, the artisanal use of colored sands to reconstruct landscapes in small bottles at Morro Branco—a typical traditional product of that area (Fig. 6.3e). The functional value is more evident where geodiversity hotspots are also identified as biodiversity hotspots: this is the case of the Ubajara National Park, in NW Ceará, where the Ibiapaba glint is a natural ramp for the settlement and staging of biological species of the Atlantic rainforest (Figs. 6.1. and 6.3f). At last, a set of active and inherited geomorphosites has a high potential for didactic uses and geoeduction in natural risks, particularly in the core of the Araripe Geopark where dormant landslides and other mass movements are largely ignored by the urban population of Crato, yet exposed to such hazards (Peulvast et al. 2011).

6.2.2 *Geoconservation: Why, What, and How?*

6.2.2.1 Threats to Geodiversity

Geoconservation deals with the conservation of abiotic nature, i.e., geological objects, landforms, and soils (Sharples 2002; Gray 2004). Because many elements of geodiversity are nonrenewable and threatened by human activities and disturbances, geoconservation is necessary for maintaining the integrity of geodiversity. In Northeast Brazil, the types of artificial activities that may degrade geodiversity depend on the types of values concerned (see above) and on the types of objects impacted, with respect to their "sensitivity." Besides the removal of geological specimens (e.g., fossils or rare minerals), mineral extraction by industrial mining and quarrying is one of the most damaging activities on landforms and landscapes. It may result in partial or total degradation of certain landforms of great scientific or other value, such as the partial destruction of a volcanic neck (Serrote Caruru: Fig. 6.4a) or that of a giant lapiaz shaped in Proterozoic dolomites (karstic

Fig. 6.3 Emblematic geomorphosites of the northern Brazilian "Nordeste." **a** The Quixadá geomorphosite (central Ceará): granitic bornhardts and spectacular inselbergs (Serra do Urucu in leucocratic facies, centre; Quixadá monzonitic complex, *left*) above the "Sertaneja" pediplain. **b** Sete Cidades National Park: karst-like and ruiniform sandstone landscapes (turrets, arches, stepped cornices…) in the core of the Cerrado ecosystem. **c** The Pontal de Santa Cruz geomorphosite (Araripe Geopark, southern Ceará): natural belvedere on the Exu sandstone cornice above the Santana do Cariri box canyon. **d** The Jericoacoara geodiversity hotspot (NW Ceará): barkhans and dune fields at the back of a quartzitic rocky coast (*background*). **e** Morro Branco, Beberibe (NE Ceará): active sea-cliffs and badlands shaped in colored sands of the Barreiras Formation overlain by palaeodune material. **f** Ubajara National Park (NW Ceará): sandstone cornice of the Ibiapaba glint with waterfall, colonized by the dense Atlantic rainforest; marble pinnacles protruding from the talus below the cornice. *Photographs* F. Bétard and J.P. Peulvast

landforms near Redenção: Fig. 6.4b). By remodeling the topography, visual impacts on landscapes are therefore one of the main consequences of mining activities. In post-mining sites, infilling of excavation pits and quarry restoration may also cause significant negative impacts, such as a loss of geological exposures and/or paleoenvironmental information through artificial burial.

Fig. 6.4 Some examples of threats to geodiversity in Northeast Brazil. **a** Serrote Caruru (northern Ceará): phonolitic neck ($\sim$35 Myear) partly degraded by mineral extraction and urbanization (*background* the city of Fortaleza). **b** The Redenção karstic geomorphosite (northern Ceará, eastern piedmont of the Baturité massif): giant lapiaz shaped in Proterozoic dolomites, partly degraded by mining activities. **c** Serra do Pereiro (eastern Ceará): dense network of parallel gullies (or '*voçorocas*') in a recently deforested area. Such gullies, of $\sim$2–3 m depth, were formed at the expense of deep red soils and friable granitic weathered materials. Deforestation is probably responsible for the accelerated soil erosion observed in that area. *Photographs* F. Bétard and J.P. Peulvast

A classical damage caused by visitors on geosites is related to vandalism and graffiti (Gray 2004). More generally, tourism pressure in sensitive sites may result in land degradation and acceleration of some erosional processes (e.g., aeolian erosion, gullying). In the coastal fringe of the Ceará state, port infrastructures and built development have caused severe perturbations to coastal dynamics (e.g., changing in long-shore drift and sedimentary transfers) and subsequent modifications of landforms (e.g., beach erosion and shoreline retreat). Technical management and coastal protection may also have negative effects because they often disrupt the operation of natural geomorphological processes. The threats to pedodiversity are of major concern as well. Many human activities (e.g., deforestation, intensive agriculture) are responsible for the degradation of soil profiles and structure, including compaction and loss of organic matter, and may result in soil acidification and accelerated erosion processes (for instance, the rapid formation of deep *voçorocas* in recently deforested hillslopes: Fig. 6.4c). Finally, more substantial threats are probably posed by the accelerated urban development in Northeast Brazil. While rapid urbanization is taking place in Fortaleza at the expense of active dunes and palaeodunes, geodiversity features can be easily destroyed due to the absence of urban or environmental planning taking into account the values of geodiversity elements. Fortunately, protected areas and geoconservation tools are effective in several areas already identified as geodiversity and/or biodiversity hotspots and in geosites of high heritage value.

6.2.2.2 Protected Areas and Geoconservation Tools

In Northeast Brazil, several elements of striking geodiversity are actually protected and managed in different ways. Facing to growing threats to the natural diversity of the Brazilian territory, the successive federal and state governments of Brazil have multiplied, for a half century, the efforts of nature conservation, by creating National Parks and reserves, and special "areas of environmental protection" (*APA*: *Areas de Proteção Ambiental*). Recently inventoried (Conto 2004; Nascimento et al. 2008), these national initiatives have been relayed locally by private initiatives, creating "special reserves of natural heritage" (*Reservas Particulares do Patrimônio Natural*) and other "ecological reserves" (*Reservas Ecológicas*) in several localities of Northeast Brazil. Historically, all the protection areas listed above were primarily employed to protect biodiversity and bioheritage, mainly in the humid mountains still covered by remnants of Atlantic rainforest. The geodiversity and geoheritage value of those areas were recognized more recently (Bétard et al. 2011), but the preexisting protection status indirectly allows the conservation of geodiversity elements, at least in strictly protected areas such as National Parks.

A first attempt to preserve specifically the geodiversity has been completed in 2006 by the creation of the first UNESCO Geopark of America and the southern Hemisphere, around the Chapada do Araripe (Herzog et al. 2008). The international recognition of the Araripe Geopark is mainly due to its palaeontological heritage, with special emphasis on the fossiliferous lacustrine/marine Albian layers of the famous Santana Formation, that contains one of the most important Cretaceous fossil associations worldwide (Martill et al. 2007; Vilas-Boas et al. 2012). Specific actions of geoconservation were developed since 2006, both in situ (quarry rehabilitation: Pedra Cariri geosite; Fig. 6.5a) and ex situ (inventory of fossil collections and renovation of the palaeontological museum of Santana do Cariri; Fig. 6.5b). The Araripe Geopark is also a great place of geomorphological heritage, although less known, with the preservation of several geosites of geomorphic interest (or geomorphosites). Located in a narrow canyon, the Ponte de Pedra geosite is a sandstone arch carved by fluvial process into the Cenomanian Exu Formation (Fig. 6.5c). This geomorphosite has a high cultural value, associated with legends and myths, considering that this natural bridge was used by Kariri Indians as a way to access a spring located in the rim of the Chapada (Olho d'Agua archeological site). Other geomorphosites of particular aesthetic value are integrated to the geosites network of the Araripe Geopark: the granitic palaeorelief of the Serra do Horto and its belvedere (at the foot of Cícero Priest statue), the Pontal da Santa Cruz geosite dominating the Santana do Cariri depression (Fig. 6.3c), and the waterfalls composing the impressive landscapes of the Batateiras and Missão Velha geosites (Fig. 6.5d). Outside the current limits of the Geopark entirely situated in the Ceará

Fig. 6.5 Geosites and potential geomorphosites of the Araripe Geopark and surroundings. **a** The Pedra do Cariri geosite: rehabilitated quarry of Albian laminated limestone (Crato member of the Santana Formation) exhibiting abundant and various fossils, as insects, fishes, algae and plants, pterosaurs, etc. **b** Palaeontological museum of Santana do Cariri: recently renovated, the museum houses one of the most comprehensive collection of Lower Cretaceous fossils in the world. **c** The Ponte de Pedra geosite: sandstone arch shaped by fluvial processes into the Cenomanian sandstones of the Exu Formation. **d** The Missão Velha waterfall: Palaeozoic sandstone (Mauriti or Cariri Formation), constituting a resistant threshold to regressive erosion (*background* Serra da Mãozinha outlier). **e** Torre Viva as a potential geomorphosite (Ipubi, Pernambuco): narrow and spectacular sandstone butte in front of the cuesta-like scarp of the Chapada do Araripe. **f** Timorante as a potential geomorphosite (Bodocó, Pernambuco): granitic tor and ruinform features in front of a glint segment of the Chapada do Araripe. *Photographs* F. Bétard and J.P. Peulvast

state, several potential geomorphosites may be proposed, such as the spectacular sandstone butte of Torre Viva (Ipubi, PE; Fig. 6.5e) or the curious granitic tor of Timorante (Bodocó, PE; Fig. 6.5f). The proposed extension of the geopark limits toward the south, in the Pernambuco state, was also considered by Barreto et al. (2012), in order to protect the fossil beds of the Santana Formation from limestone and gypsum mining, illegal collecting and inadequate legislation.

6.2.3 Geotourism and Geoeducation: New Challenges in Northeast Brazil

6.2.3.1 Geotourism and Sustainable Local Development

Geotourism is emerging as a new form of cultural tourism and as a branch of sustainable tourism (Dowling 2011; Farsani et al. 2011). Focusing on abiotic nature and geodiversity, geotourism adds opportunities to cultural discovery and rural development, by enhancing visitors and tourists to travel in a territory in order to experience, learn from, and enjoy geoheritage. Geotourism is an integral part of UNESCO's geopark strategies and is essential to their maintenance and development. In Northeast Brazil, the Araripe Geopark in southern Ceará has experienced a process of geotourism development since its creation in 2006. This process involves both the local community and the visitors, all of whom benefit in their own way from the promotion of geoheritage. Through geotourism, it is stated that the geopark should aim for the development of a regional territory and support local economy. By attracting increasing numbers of visitors, the geopark effectively stimulates local socioeconomic development through the promotion of an internationally recognized label, that encourages the creation or development of local enterprises and cottage industries involved in geotourism and geoproducts. However, in the Araripe region, this activity must stay compliant with the maintenance, in several parts of the geopark, of controlled and environmentally responsible mineral extraction, since this traditional activity is also important for local economy (Barbalha, Crato, Nova Olinda, Santana do Cariri: extraction of clays for ceramic industries, quarrying of laminated limestone and gypsum for construction geomaterials).

Apart from the Araripe Geopark, where the durability of geotourism is one of the main challenges in the next years, other potential destinations of geotourism might be promoted in the Ceará state and adjacent regions. The Sete Cidades and Ubajara National Parks, respectively at the interior and the border of the Parnaiba basin, are privileged geotouristic destinations for international people, although the two parks are mainly known as places of ecotourism promoting the aspects of biodiversity and of particular ecosystems. The Quixadá geosite, in central Ceará, is an incontournable place for geotourism activities as well. Finally, the chaplet of coastal geomorphosites (Icapui, Beberibe-Morro Branco, Jericoacoara, Camocim) may formalize a geo-road easily accessible from Fortaleza. Geotourism is undoubtedly a new challenge in Northeast Brazil and will be supported by the exceptional geodiversity of its territory.

6.2.3.2 Geoeducation and Communication on Georisks

Geoeducation is one of the objectives of UNESCO's geopark strategies. In the Araripe Geopark, many geoeducational activities are organized for scholars but also for the general public (seminars, school class excursions, educational workshops…), mainly by the way of local community awareness on the importance of protecting the environment and the palaeontological heritage (e.g., actions to prevent illegal collecting of fossils). Interpretative centers at the Crato office and the palaeontological museum of Santana do Cariri provide logistic support to communicate geoscientific knowledge to the public. Concentrating its communication on the preservation of the palaeontological heritage, the Geopark authorities are encouraged to develop other geoeducational purposes and to communicate to a wider community of people. Because the Araripe Geopark is also a high place of bioheritage and biodiversity—with the Araripe National Forest, or FLONA, established since 1946 and the APA Chapada do Araripe—, the links between geodiversity and biodiversity need to be better emphasized in the future, in order to facilitate a greater appreciation of natural diversity and its complexity.

A major challenge in geoeducation inside the Araripe Geopark and, more broadly, in Northeast Brazil, is associated to communication on natural hazards and risks. The notion of natural risk remains poorly present in the mentalities and behaviors, as well among the inhabitants as the local actors, and it does not appear in management and urban planning, except in the form of more or less isolated technical measures responding to successive crises (Magalhães and Peulvast 2013; Magalhães 2014). In the Araripe Geopark, the catastrophic floods of the Grangeiro River, which seriously and regularly damaged the city of Crato, are a good example of the sudden events, which threaten properties, infrastructures and lives in Northeast Brazil. Being more frequent than the large-scale mass movements recently discovered in the rim of the Chapada do Araripe (Peulvast et al. 2011), the floods, particularly flash floods, remain poorly studied and treated, in spite of the numerous political announces repeated after each flood. Because landslides and floods are natural hazards related to geomorphological and hydrological processes, a new mission of the Araripe Geopark might be to develop geoeducational actions on georisks that directly threaten the local population. In our opinion, this is a powerful way of improving communications on geosciences and on the importance of taking into account landform and landscape evolution processes at various temporal and spatial scales.

References

Barreto AMF, Brilha JBR, Sales AMF, Almeida JAC (2012) Patrimônio paleontológico e geoconservação da Formação Santana (Cretáceo Inferior da Bacia do Araripe, Pernambuco e Piauí: Nordeste do Brasil). In: Henriques MH, Andrade AI, Quinta-Ferreira M, Lopes FC, Barata MT, Reis RP, Machado A (eds) Para aprender com a Terra: memórias e notícias de Geociências no espaço lusófono. Imprensa da Universidade, Coimbra, pp 311–320. doi:10.14195/978-989-26-0533-3_33

Beard JS (2003) Palaeodrainage and the geomorphic evolution of passive margins in southwestern Australia. Palaeodrainage and the geomorphic evolution of passive margins in southwestern Australia (with 7 figures). Zeitschrift für. Geomorphologie 47(3):273–288

Bétard F (2007) Montagnes humides au cœur du Nordeste brésilien semi-aride: le cas du massif de Baturité (Ceará). Apports d'une approche morphopédologique pour la connaissance et la gestion des milieux. Ph.D. thesis, Univ Paris-Sorbonne
Bétard F (2010) Uplift and denudation history at low-elevation passive margins: insights from morphostratigraphic analysis in the SE Armorican Massif along the French Atlantic margin. CR Geoscience 342:215–222. doi:10.1016/j.crte.2010.02.004
Bétard F (2012) Spatial variations of soil weathering processes in a tropical mountain environment: the Baturité massif and its piedmont (Ceará, NE Brazil). Catena 93 (18–28):2012. doi:10.1016/j.catena.01.013
Bétard F, Peulvast JP (2011) Evolução morfoestrutural e morfopedológica do maciço de Baturité e de seu piemonte: do Cretáceo ao Presente. In: Bastos FH (ed) Serra de Baturité: Uma Visão Integrada das Questões Ambientais. Expressão Gráfica e Editora, Fortaleza, pp 35–59
Bétard F, Peulvast JP (2013) Non-lateritic red soils of the Brazilian drylands: pedogenic processes and landscape evolution. In: 8th International conference on geomorphology. Paris, 27–31 Aug 2013, Abstracts volume, p 594
Bétard F, Peulvast JP, Claudino Sales V (2005) Laterite preservation and soil distribution in the Araripe-Campos Sales area, Northeastern Brazil: consequences of uplift, erosion and climatic change. In: 6th International conference on geomorphology. Zaragoza, 7–11 Sept 2005, Abstracts volume, p 69
Bétard F, Peulvast JP, Claudino Sales V (2007) Caracterização morfopedológica de uma serra úmida no Semi-árido do Nordeste brasileiro: o caso do maciço de Baturité-CE. Mercator 12:107–125
Bétard F, Caner L, Gunnell Y (2009) Bourgeon G (2009) Illite neoformation in plagioclase during weathering: evidence from semi-arid Northeast Brazil. Geoderma 152:53–62. doi:10.1016/j.geoderma.05.016
Bétard F, Peulvast JP, Rabassa J, Aguilera E (2014) Meso-Cenozoic paleotopographies and paleolandscapes in the Deseado Massif (Santa Cruz province, Argentina). In: Rabassa J, Ollier C (eds) Gondwana landscapes in southern South America. Springer, Berlin, pp 477–501
Brown RW, Gallagher K, Gleadow AJW, Summerfield MA (2000) Morphotectonic evolution of the South Atlantic margins of Africa and South America. In: Summerfield MA (ed) Geomorphology and global tectonics. Wiley, Chichester, pp 255–281
Brunsden D (1993a) The persistance of landforms. Z Geomorph Suppl Bd 93:13–28
Brunsden D (1993b) Barriers to geomorphological change. In: Thomas DSG, Allison RJ (eds) Landscape sensitivity. Wiley, Chichester, pp 7–12
Conto RG (2004) Atlas de conservação da natureza brasileira. Unidades federais, Metalivros, São Paulo
CPRM-Serviço Geológico do Brasil (2006) Mapa geodiversidade do Brasil. CPRM, Brasília
Dowling RK (2011) Geotourism's global growth. Geoheritage 3(1):1–13
Farsani NT, Coelho C, Costa C (2011) Geotourism and geoparks as novel strategies for socio-economic development in rural areas. Int J Tourism Res 13(1):68–81
Gray M (2004) Geodiversity: valuing and conserving abiotic nature. Wiley, Chichester
Gray M (2008) Geoheritage 1. geodiversity: a new paradigm for valuing and conserving geoheritage. Geosci Can 35(2):51–59
Guichard E (1970) Les sols du Bassin du Rio Jaguaribe. Mémoire ORSTOM 40, Paris
Gurgel SPP, Bezerra FHR, Corréa ACB, Marques FO, Maia RP (2013) Cenozoic uplift and erosion of structural landforms in NE Brazil. Geomorphology 186:68–84
Harman R, Gallagher K, Brown R, Raza A (1998) Accelerated denudation and tectonic/geomorphic reactivation of the cratons of northeastern Brazil during the Late Cretaceous. J Geophys Res 103:27091–27105
Harris SE, Mix AC (2002) Climate and tectonic influences on continental erosion of tropical South America, 0–13 Ma. Geology 30(5):447–450
Herzog H, Sales A, Hillmer G (2008) The UNESCO araripe geopark: a short story of the evolution of life, rocks and continents. Expressão Gráfica e Editora, Fortaleza

Ibañez JJ, De-Albs S, Bermúdez FF, García-Álvarez A (1995) Pedodiversity: concepts and measures. Catena 24(3):215–232

Japsen P, Bonow JM, Green PF, Cobbold PR, Chiossi D, Lilletveit R, Magnavita LP, Pedreira A (2012) Episodic burial and exhumation in NE Brazil after opening of the South Atlantic. Geol Soc Am Bull 124:800–816

Kooi H, Beaumont C (1994) Escarpment evolution on high-elevation rifted margins: insights derived from a surface processes model that combines diffusion, advection, and reaction. J Geophys Res 99:12191–12209

Mabesoone JM, Castro C (1975) Desenvolvimento geomorfológico do nordeste brasileiro. Bol Núcleo Nordeste da SGB, Recife 3:5–35

Magalhães AO (2014) Crues éclair et inondations urbaines dans le sertão du Nordeste brésilien (Rio Grangeiro, Crato, Ceará): perception et gestion d'un risque naturel méconnu. BAGF-Géographies 3:257–275

Magalhães AO, Peulvast JP (2013) Aléas et risques naturels dans l'intérieur semi-ride du Nordeste brésilien: la crue-éclair du rio Grangeiro (Crato, Ceará) du 28 janvier 2011. EchoGéo 26. doi:10.4000/echogeo.13596

Magnavita LP, Davison I, Kuznir NJ (1994) Rifting, erosion and uplift history of the Reconcavo-Tucano-Jatoba Rift, Northeast Brazil. Tectonics 13:367–388

Marques FO, Nogueira FCC, Bezerra FHR, de Castro DL (2014) The Araripe basin in NE Brazil: an intracontinental graben inverted to a high-standing horst. Tectonophysics 630:251–264

Martill DM, Bechly G, Loveridge RF (2007) The crato fossil beds of Brazil: window into an ancient world. Cambridge University Press, Cambridge

Migoń P (2006) Granite landscapes of the world. Oxford University Press, Oxford

Mochiutti NF, Guimarães GB, Moreira JC, Lima FL, Freitas FI (2012) Os Valores da Geodiversidade: Geossítios do Geopark Araripe/CE. Anuário do Instituto de Geociências – UFRJ 35(1):173–189

Morais Neto JM, Hegarty K, Karner GD (2005–2006) Abordagem preliminar sobre paleotemperatura e evolução do relevo da bacia do Araripe, Nordeste do Brasil, a partir da análise de traços de fissão em apatita. Bol Geoci Petrobras 14:113–119

Morais Neto JM, Vasconcelos PM, Stone J, Lima MG (2010) Variability of erosion rates captured by ^{10}Be cosmogenic isotope: a case study in the Seridó valley, northeastern Brazil. VII SSAGI-South American Symposium on Isotope Geology, Brasilia

Nascimento MAL, Ruchkys UA, Mantesso-Neto V (2008) Geodiversidade, geoconservação e geoturismo. Trinômia importante para a proteção do patrimônio geológico. Soc Bras Geol, São Paulo

Panizza M (2009) The geomorphodiversity of the Dolomites (Italy): a key of geoheritage assessment. Geoheritage 1:33–42. doi:10.1007/s12371-009-0003-z

Pazzaglia FJ, Gardner TW (2000) Late Cenozoic landscape evolution of the US Atlantic passive margin: insights into a North American great escarpment. In: Summerfield MA (ed) Geomorphology and global tectonics. Wiley, Chichester, pp 283–301

Pereira DI, Santos L, Silva J, Pereira P, Brilha JB, Silva J, Rodrigues C (2012) Mapping regional geodiversity in Brazil and Portugal. Geologia dell'Ambiente, Suppl 3:107–108

Peulvast JP, Bétard F (2013) Late Cenozoic and present-day hillslope erosion dynamics in a passive margin context: stability or instability? Case studies in Northeast Brazil. Geografia Fisica e Dinamica Quaternaria 36(1):139–149. doi:10.4461/GFDQ.2013.36.0

Peulvast JP, Bétard F (2015) A history of basin inversion, scarp retreat and shallow denudation: the Araripe basin as a keystone for understanding long-term landscape evolution in NE Brazil. Geomorphology 233:20–40. doi:10.1016/j.geomorph.2014.10.009

Peulvast JP, Claudino Sales V (2004) Stepped surfaces and palaeolandforms in the northern Brazilian "Nordeste": constraints on models of morphotectonic evolution. Geomorphology 62:89–122

Peulvast JP, Vanney JR (2002) Géomorphologie structurale, tome 2. Relief et géodynamique. CPI-GB Science Publication, BRGM Editions, Paris, Orléans

Peulvast JP, Claudino Sales V, Bétard F, Gunnell Y (2008) Low post-Cenomanian denudation depths across the Brazilian Northeast: implications for long-term landscape evolution at a transform continental margin. Global Planet Change 62:39–60. doi:10.1016/j.gloplacha.2007.11.005

Peulvast JP, Bétard F, Magalhães AO (2011) Scarp morphology and identification of large-scale mass movements in tropical tablelands: the eastern Araripe basin (Ceará, Brazil). Géomorphologie: relief, processus, environnement 1:33–52. doi:10.4000/geomorphologie.8800

Pugh JC (1966) The landforms of low latitudes. In: Drury GH (ed) Essays in geomorphology, Heinemann edn. Books Ltd, London, pp 121–138

RadamBrasil P (1981a) Levantamento integrado dos recursos naturais do Brasil. Folha Fortaleza. Ministério das Minas e Energia-MME, Brasilia

RadamBrasil Projeto (1981b) Levantamento integrado dos recursos naturais do Brasil. Folha Jaguaribe-Natal, Ministério das Minas e Energia-MME, Brasilia

Sharples C (2002) Concepts and principles of geoconservation. Tasmanian Parks and Wildlife Service website

Silva CR (2008) Geodiversidade do Brasil: conhecer o passado, para entender o presente e prever o futuro. CPRM, Serviço Geológico do Brasil

Tennesen (2014) Rare earth. Science 346:692–695

Thomas MF (1989a) The role of etch processes in landform development. I. Etching concepts and their applications. Z Geomorph NF 33(2):129–142

Thomas MF (1989b) The role of etch processes in landform development. II. Etching and the formation of relief. Z Geomorph NF 33(3):257–274

Vilas-Boas M, Lima F, Brilha JB (2012) Conservation of the palaeontological heritage of Araripe Geopark (Ceará, Brazil): threats and possible solutions. Geologia dell'Ambiante Suppl 3:87–88

Index

J.-P. Peulvast and F. Bétard, *Landforms and Landscape Evolution of the Equatorial Margin of Northeast Brazil*, Springer Earth System Sciences,
DOI 10.1007/978-3-319-18203-2

FSC
www.fsc.org
MIX
Papier aus verantwortungsvollen Quellen
Paper from responsible sources
FSC® C105338

If you have any concerns about our products,
you can contact us on
ProductSafety@springernature.com

In case Publisher is established outside the EU,
the EU authorized representative is:
Springer Nature Customer Service Center GmbH
Europaplatz 3, 69115 Heidelberg, Germany

Printed by Libri Plureos GmbH
in Hamburg, Germany